LEARN ARDUINO BEGINNERS TO ADVANCE

Complete Self Guide, Example code, Motors, Sensors, Syntax, Comparisons

BY

Sathish Kumar

ABOUT THE BOOK

Did you know that one of the main elements that form the internal structure microcontrollers are the registers? Every time you create code for the Arduino, at some point you are manipulating the internal registers of your microcontroller. But after all, Do you know they are the registers? In today's video, Silicon Lab and PCB Way will get you present the importance of existing registrars in the Microcontroller of your Arduino and yet we will offer you 10 free printed circuit board. To do this, just click on the link of the description, make your account and earn 10 free prototypes. But now, let's start another video of series: Inside Arduino and getting to know the registrars. The registers are elements that have the functioning similar to that of memory, i.e. they serve to store and register value. But that is not its main function. Is through this stored information that we will be able to configure the operation of the internal circuits of our microcontroller. Maybe you have I was in doubt. And now I will show you An example. Imagine you have a register that controls its two modes of speed: walking and running and their two Mood types: sad and happy. That is, through it, we will control its two characteristics. The process is simple. Imagine that the register is like a cashier. We have two types of speed and two types of mood, we will only use a one-bit box for each one. I.e. for the speed we will use 0 for walking and 1 for running. And for mood, we will use 0 for sad and 1 to happy. Now, let's understand the process of configuration. First, let's set up so that you walk happy. A what should we do? In the first position, we will put 0 and in the second position we will put 1, that is, with this mode, you will walking happily. And if you want to be set up to run happy? Is very simple! We need to put in the position of speed as 1 and mood position as 1. Did you see how important it is to understand the operation of registers? All microcontroller has a multitude of registers. And among them, we highlight the registers for circuit configuration, registers used to signal CPU internal states, registers to configure input pins output and others in your microcontroller. But for ease, let's look at the pin registers of I/O. There are three types of configuration registers, which are: the direction registers, output registers and the input of the register. The direction registers are used to configure a certain pin as input or output. The output registers are used to drive a pin when It is configured by the register direction as output. While the input register will be used to read the state of a pin when it is configured as an input. Let's look at the process below. All pins microcontrollers are divided into groups of eight. These groups we call the port, which is defined by an 8-bit register. Imagine then that you want to activate pin 4. And what should we do? First, you must configure that pin as output and then make your activation. To do this we must place 1 in position D4 to indicate the configuration as output and 0 in the others, to be configured with inputs. But be careful! Verifique simper o datasheet, para sabre quall the logical level used to configure the pin as input and output. After this process, it's time to activate the D4 output. We will use the other register for the outputs. In this case, we will put 1 in position D4 and 0 in the others, to be deactivated. That way we can only activate pin D4. Did you see how the process is simple? Yeah, we often use the Arduino functions and we didn't realize that we are manipulating several internal configuration registers. Because the function facilitates and already makes all the setting by ourselves. But after all ... Why do we use programming of registers? First, because it reduces the size of programming code, has greater speed access for activation and pin reading and access speed to several registers.

TABLE OF CONTENTS

ABOUT THE BOOK 2

WHAT IS ARDUINO 7

EARN WITH ARDUINO 14

THE HOLES IN BOARD 16

RASPBERRY PI CONNECTIONS 17

RASPBERRY PI CONFIGURATION AND CODE 19

ARDUINO RASPBERRY PI CONNNECTION 21

FINISHING CONFIGURATION 22

THE BREADBOARD 24

WHAT IS A MICROCONTROLLER 26

HOW DOES ARDUINO THINKS 27

ARDUINO SOFTWARE FOR THE FIRST TIMERS 40

ARDUINO SYNTAX 51

ARDUINO SYNTAX WITH EXAMPLE 56

PUTTING TOGETHER A CIRCUIT 62

WHAT IS ARDUINO RASPBERRY PI AND PIC MICROCONTROLLER 65

QUICK SUMMARY OF THE MAIN DIFFERENCES 68

HARDWARE POWER AND 69

CONNECTIVITY-EN 69

SOFTWARE COMPARISON 72

EXPANDING ONWARD CAPABLITIES 73

BASIC PARTS REQUIREMENTS 74

PARTS NEEDED TO GET THE JOB DONE 75

WIRE UP A POWER SUPPLY 76

ARDUINO READY	78
BOOTLOADING YOUR CHIPS OPTIONAL	80
FINISH UP AND BURN THE BOOTLOADER	83
ATMEL ATMEGA 328 DETAILS	85
ARDUINO SOFTWARE FOR THE FIRST TIMERS	87
DIGITAL READ AND SERIAL PORT	89
IF CONDITIONAL STATEMENT	91
FOR LOOP	98
WHILE LOOP	104
USING ARRAYS	110
DIFFERENCE BETWEEN INPUT AND OUTPUT	116
MULTIPLE LED	118
BUTTON INPUT AND LED	123
INPUTS BUTTONS	127
ANALOG INPUT IR SENSOR	129
ANALOG INPUT POTENTIOMETER	131
ANALOGUE INPUT WITH EXAMPLE	136
LIQUID CRYSTAL DISPLAY	139
WIRELESS POWER INTRODUCTION AND HISTORY	145
HOW WIRELESS POWER WORKS	146
ADVANTAGES AND DISADVANTAGES	148
RANGE OF WITRICITY AND WITRICITY APPLICATIONS	149
TYPES OF MOTORS	151
SIMPLE DC MOTOR	155
TYPES OF MOTORS	158
SERVO MOTOR	162
STEPPER MOTOR	168

RUNNING A DC MOTOR	175
RUNNING A SERVO MOTOR	180
ARDUINO BLUETOOTH	183
HOW DOES IT WORK	184
ANDROID APPLICATION	186
SCHEMATIC	188
CODE	188
TESTING CODE	191
INTRODUCTION TO SENSORS	193
POTENTIOMETER	199
PIEZOELECTRIC SENSOR	205
TEMPERATURE SENSOR	210
PASSIVE INFRARED SENSOR	216
SERIAL COMMUNICATION INTRODUCTION	219
WHAT IS ULTRASONIC SENSOR	225
SOFTWARE REQUIREMENTS SCHEMATIC	226
TEMPERATURE AND HUMIDITY SENSOR	228

WHAT IS ARDUINO

You are going to get me of the ideas about the basics of Arduino. And in the second one me of the examples about how the programming can be done with Arduino. before we proceed further, I would like to remind you a few things that Arduino is very much popular at present, it is used in four different implementations of IoT throughout the world. Arduino devices are very much cheap, they are low reduce consuming and that is why they are very much popular for us in the implementation of the internet of things. in the first module, we have seen different things we have understood the concepts of the Internet of Things, basic concepts, the overall features of the philophy of the Internet of Things. We have al seen that there are different types of senrs, different types of sensing possible different types of actuators, the principles behind different types of sensing different types of actuation. We have seen that there are different types of networks that are possible for us. Production for use in IoT different types of communication devices standards can al be used for communicating in the Internet of Things. having understood those, how can we use these concepts for building a real internet of things may be in a smart home scenario at home to improve me of them you know, daily tasks that we do at home or in smart homes, smart city scenario, like in a smart hospital, smart city, you know, smart transportation, connected vehicles and on. for all these, we need to take help of different IoT devices. And one of the popular ones is Arduino. our do we know if we have to use for the building of Internet of Things. You have to buy these which are pretty cheap.

And then you have to program these, and this is what I'm going to teach you in this particular course. with me, I have Mr. unimproved Mukherjee who is going to take over and Mr Mukherjee is going to take you through the hands-on of Arduino programming to the from the starting from the basics to it will be moderately advanced concepts of Arduino programming. I would like to show you first how an Arduino device looks like. this is the Arduino Uno. Arduino Uno, Arduino has different variants, they all have different you know, differences in specifications and on. this is the Arduino Uno. And this is this device that has to be programmed. This is this has to be programmed, as you can see over here, this is very small in size and it can be very much integrated with these Internet of Things. You know, when you're trying to implement the Internet of Things. It can be implemented on top. this device actually, you know the different senrs that you have you know learnt in this particular course. these different senrs can be fitted to this device the different actuators can be fitted to this device and these senrs, the different senrs and the different actuators in after fitting the data that is received from the same urce. These can be sent to the communication unit in this which on the rupee is going to talk about and how this data can be disseminated and can be sent for further analytics for storage and on. I'll be taking you through the basic features of Arduino in this lecture. to start with, as you already have heard, Arduino has become very popular nowadays. first of all, the main reans are, it is an open-urce programmable board

Features of Arduino

- Open source based electronic programmable board (micro controller)and software(IDE)
- Accepts analog and digital signals as input and gives desired output
- No extra hardware required to load a program into the controller board

with built-in microcontrollers, and a ftware ID and this ftware ID will help you change the behaviour of the microcontroller according to your needs. it accepts analogue as well as digital signals which can be given as input and it will give outputs which are mainly digital. no extra hardware is required to load a program into the controller vote. for the people who have worked with 8051 series, microcontrollers, 8085 microprocesr, they must have remembered that you needed an extra programmer to program the procesr board and there were lots of interfacing ICS and all those things are not required with the Arduino based systems.

Types of Arduino Board

- Arduino boards based on ATMEGA328 microcontroller

- Arduino boards based on ATMEGA32u4 microcontroller

- Arduino boards based on ATMEGA2560 microcontroller

- Arduino boards based on AT91SAM3X8E microcontroller

to start with, there are a few basic variations of the Arduino boards. They have there are ATMEGA 328 based microcontrollers, they have ATMEGA 32 series microcontrollers, they have ATMEGA 2560 series microcontrollers and then there are ATMEGA 91 Sam, three x 80 series microcontrollers.

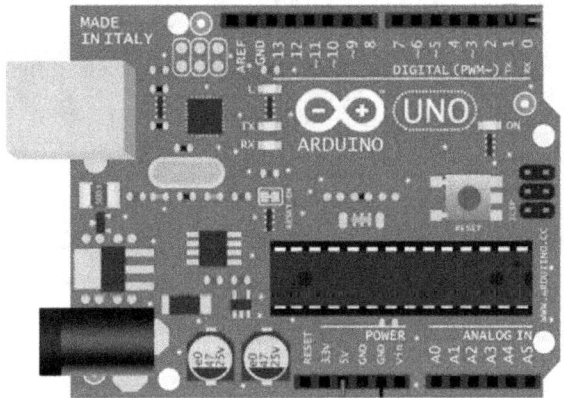

these are me of the core microcontrollers as you can see if you focus on this IC this Arduino board, we are using uno board to give a demonstration. this is the IC chip and basically, all the other ones are either the voltage converters or interfacing ICS which are required for the input-output functions with this ATMEGA series chip.

Arduino UNO

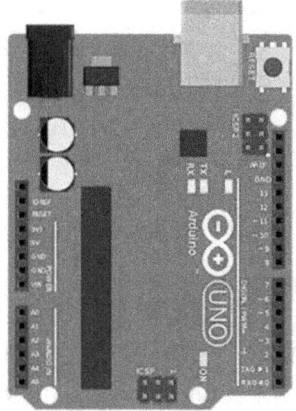

Feature	Value
Operating Voltage	5V
Clock Speed	16MHz
Digital I/O	14
Analog Input	6
PWM	6
UART	1
Interface	USB via ATMega16U2

basic features of Arduino Uno they operate At a voltage of five volts with a clock speed of 16 megahertz and they have 14 normally Arduino Uno has 14 digital input-output pins, six analogue input pins, six PW m pins, one party that is the universal asynchronous receiver and transmitter and the interface is mainly via USB of ATMEGA 16 you to.

Board Details

- Power Supply: USB or power barrel jack
- Voltage Regulator
- LED Power Indicator
- Tx-Rx LED Indicator
- Output power, Ground
- Analog Input Pins
- Digital I/O Pins

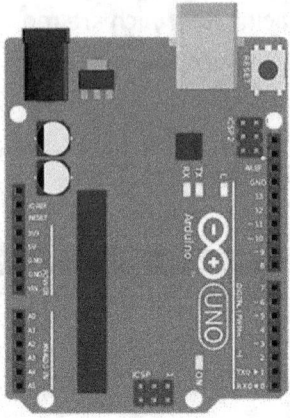

the more details are as you can see, from this figure, you have a USB connector to which you connect your ATMEGA boat to your PC. The best thing about this is the interfacing is very easy. You can connect your system Arduino based mode to either a Windows-based PC or a Macintosh or a Windows or Linux based system. this is the USB connector. Then you have the power connector to power on the device in standalone mode. Otherwise, if you connect it to a PC it draws power from the PC itself. And this is the analogue reference pin. You have 14 digital pins, which can be used as input and output. As you can see starting from zero to 13. These are the 14 input-output pins. And over here you have six analogue pins, a zero to a five which can receive analogue inputs. And these are me of the power connectors. You have a five volt 3.3 volt and ground connections and on. these are just me of the basic components of the Arduino.

Arduino IDE

- Arduino IDE is an open source software that is used to program the Arduino controller board
- Based on variations of the C and C++ programming language
- It can be downloaded from Arduino's official installed into PC

Now, the Arduino ID is open-urce ftware. The Arduino system itself is an open-urce system the hardware specifications are available. You can in fact if you have fabrication facilities, you can fabricate your own Arduino device. this ID no ID is an open-urce urce When that is used to program the Arduino board, it is based on the variations of C and c++ programming languages, and it can be freely downloaded from the Arduino official website.

EARN WITH ARDUINO

How to make money with Arduino. Robotics is a dreamy field inventing the future armor of iron man or even making more reasonable projects that you can use in our everyday life through the spirit of an inventor Robotics is a world full of promise and gives us the power to remove problems with new solutions It may sound complicated and we think that you have to study out for a long time on that we need significant finances to make it happen If you wanted to realize a robotic project to commercialize it you need electronic components tools hardware-software collaborators means of production money and therefore investors that only a company can afford This is why we are entitled to wonder if this dream is accessible or if it is too big But when we want to succeed we cannot give up. I just want to make my robotic project I will do a project every week. I have butterflies in my stomach Today with any computer on just an Arduino board on components like sensors on actuators You can realize promising robotic project and thanks to the Internet You can make your invention known and start selling it on a crowdfunding platform among the best-known are Kickstarter and Indiegogo If you have a good product idea and a prototype Just make a great sales video to raise funds by putting your project in pre-sale to a community of new technology enthusiasts present on this website. Kickstarter allows you to make your project visible in exchange for 5% Some manage to raise hundreds of thousands of dollars and others millions of dollars However, it is not so simple. The success rate is 36% and three-quarters of those who fail to get less than 20%. The least we can say is that either it likes or not at all. After you Kickstarter campaign, you will have the funds to start the production and this will have allowed you to make yourself known As you are a small team and perhaps still alone, you do not need to get a lot of customers to make your business viable You can see successful campaigns and yet the concept is simple. The key is that he still responds to a real need. The testimonies of those who have launched projects are numerous many fails which is why you need to know the key points that will make you project a success. These successes are not due to mysterious reasons. They are simply logical. For your summary a few keys I will tell you first of all that a video of sale must respect some codes like storytelling or the art of telling the story of your project Copywriting or the art of persuading that your product is the right one. (with content that appeals to emotions visualization ND to your 5 senses). Your sales video must be so understandable that even if you do not listen to the visual will be enough to understand everything The proof of authority that must show your mastery of this subject. Highlighting the benefits of your product This is a very important point. Be present on social networks several months if possible before your launch. This is the social proof preparing for the launch of your Kickstarter campaign by retrieving emails from interested people that you got via social networks and/or your website. And this to warn them that your campaign will be launched. Indeed Kickstarter highlights projects that have funding coming in the early days. Contact bloggers, influencers and media in your industry to help you market your product. And of course, having motivation on patience. So choose projects that you like to stay on course knowing that you will not be rewarded until you finalize your campaign Maybe you have no idea in this case There are methods to find an idea or at least to identify if an idea is good. The first question is whether your product meets a need and therefore solves a problem that is important to you and If you will buy it for yourself to solve the problem. The goal of this YouTube channel is to learn to invent a product but also to sell them. We will learn to realize projects by taming the use of electronic parts and by carrying out more elaborate projects. Projects aiming at an entertainment or to facilitate our lives but also in the purpose of inventing a product to market.

THE HOLES IN BOARD

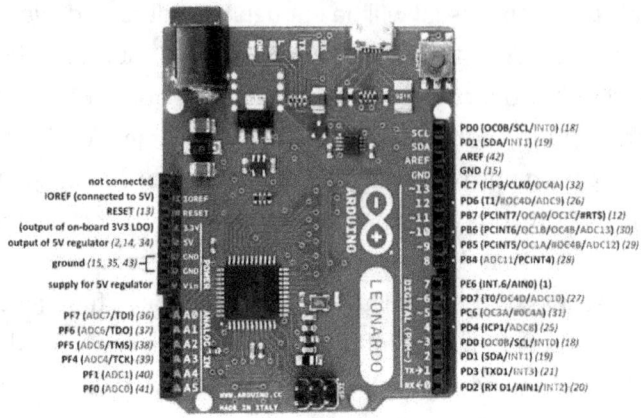

We will talk about the holes in the board saw in the previous picture there are lines of holes running up and down either side of the board. You will be sticking things in those holes. There are two other holes you may be sticking things in and that's the two in the front of the board. As you can see two places where you can plug things in Generally you will use the left one to attach USB plug and thereby to your computer and you will use the right one.to power your board when it's away from your computer by using either a wall plug or batteries the left plug will be used to upload the code into the board. It can be used to control whatever hardware you attach to the little holes on the left and the right That's what an Arduino is. Thanks for watching this lesson. If you have any questions please ask it in the Q and A board. Now you know what an Arduino is. And what does the next step is to show you how it does it. Because there is a thriving development community for Arduino on the Internet. You can probably find a code and a circuit diagram for any project you want to make. However, that requires you to be able to read code and understand it That will be covered later in this course. Thanks for watching this Lesson. If you have any question please ask it in that Q and A board. See you next lesson.

RASPBERRY PI CONNECTIONS

This is the cable that we will use to connect the Arduino to the rasberry body. So let's get started. For the connection box to connect the lead to number 11 as shown in the picture below. This is a breadboard. This is rasberry by. And this is a lead. We will connect this led to being number 11 in the army in a ball. The other treadmill will be well connected to ground.

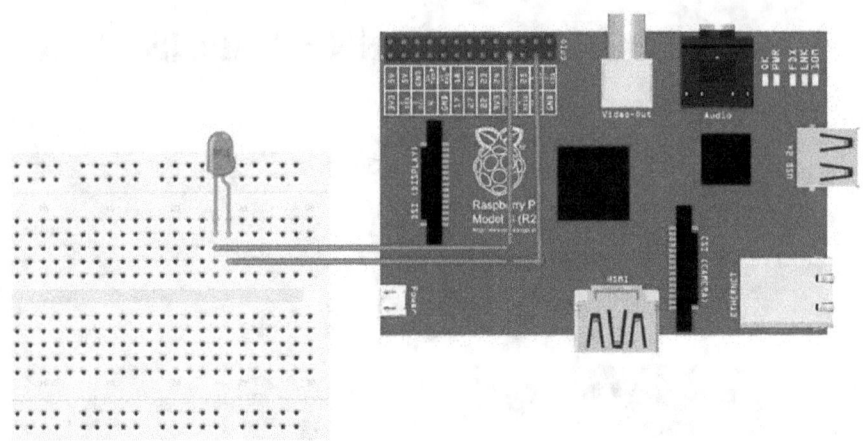

So let's start by using the fluty izing software for creating this simple circuit. Now here we have a software for visiting circuit simulation and circuit design. It's called Flight zing. It's FREE. You can go get it and download it go to the ceviche. I trust billyboy. And you will find that we have respited by boughts. Let's go with the hospital by B. All the fears will of trust billyboy. You can't choose any of these. As you can see OK this is the hospital body would be. Now we need a good. This is OK. This is Eilidh. We also need a big board OK. Now we have our board we need all this led to this ball. But before going any further we need to review this connection. As you can see this spin down then must be connected to the ground of the Arduino board. The other bin which is this one must be connected in number 11 OK. Now if it's on here you will find that this been our number this is number 9 8 11. This is the number 11 G-B or 11 as you can see. So we go here and we can connect it with this OK. That's it. We made the last bit about connection as you can see here. We've connected the ground. There's been a number of them of the last bit about this lead. This lead will blink when it receives a sentence from that we know.

RASPBERRY PI CONFIGURATION AND CODE

```python
import serial
import RPi.GPIO as GPIO
import time

ser=serial.Serial("/dev/ttyACM0",9600)  #change ACM number as found from ls /dev/tty/ACM*
ser.baudrate=9600
def blink(pin):

    GPIO.output(pin,GPIO.HIGH)
    time.sleep(1)
    GPIO.output(pin,GPIO.LOW)
    time.sleep(1)
    return

GPIO.setmode(GPIO.BOARD)
GPIO.setup(11, GPIO.OUT)
while True:

    read_ser=ser.readline()
    print(read_ser)
    if(read_ser=="Hello From Arduino!"):
        blink(11)
```

Let's start by turn and go in the right spirit by and ORBIN by three in new window if you will. Yet talk about trust barabar you can check our course or waspishly by step by step guide to learn how to turn it on how to use it and how to write your own code inside it. You won't need to write the following code in the new window and save it saving it to your desktop is a good place so you don't lose it. This is the code. Let's start by explaining this very simple rasberry by code so that you will know what you are writing. First we are importing serial communication and time libraries. We are also importing RB bearable input output library which is for us by Ben's input and output barbus input output. This is what this stands for. These are for importing. Next we are defining the serial communication so it equals serial. Serial. This is the location for the serial communication device which is in this case the Arduino. We will change this and that fuel coming lessons but for now you need to understand this line and this is the Badreya 9600 changing. It's the I.M. but which is here as found from us. Dev D d y se. If we are this command we will see a list of the serial communication device after connecting the dots Beary we all find something like this with an iron bar after it. We will need to change that number with the new bar after connectome. We will talk about this again in the few coming lessons. C of the quadrate is 9600. This is for sitting at about the rate of the serial communication between the Arduino and the rasberry by next we have the diff blink been. This is for blinking LED Now John Berberis and would opt out. But for Ben which is he is just a bit of input output high so it will send 5 volts or 1 logic to this been so that it can blink it then it will wait for one second and then it will send a low which is zero. Then it will wait for one minute second after that it will. Again to hear. If so this is like a veteran beat repeats itself. That means returning to the point of the cord. When this function was called after that we have general purpose and sit mode for connecting the board and there was an optimal setup for bin number 11 to which we connected the LED we defined this Ben as out but just a bit of put out so that it can consider considered as output. Now while true while true we needed to read the serial line lead C which is c of the tree line it will read the incoming data using the area of communication between us better buy and store these values or this data to the value of valuable read said this research will be Brinton's. But in the lead set in the window inside rasberry by now if the value inside read search equals Hello from Arduino and this mark it will Blinkx number 11. Number 11 and Blinkx will go here. To execute this. Then returning it. So as you can see blank is a function inside bifold language. This was sublimation of the rasberry bytecode. Let's revise it these for importing the library's serial time and input output libraries. These are for these two lines for defining in communication protocol and defining which US be serial communication been is the one that Albinus connected to. This link to the data. These lines are a function called blink when we define that ban it will use it to blink that Ben these two lines for defining the number 11 as output for defining the board. Now this one will keep repeating itself while through it will read the scene communication data that are being sent to the rasberry it will print it. Then it will ask if it equals this line it will Blinkx number 11 blink means call the function call the blank and execute its command on bin number 11 so it will go here. It will start as high then low which means that then it will turn on and off then it will have to wait for fear of instruction. That's it for the last bit of bytecode and rasberry by configuration. I know that I explained the code very fast but since it's really simple and we don't want to waste time on this code.

ARDUINO RASPBERRY PI CONNNECTION

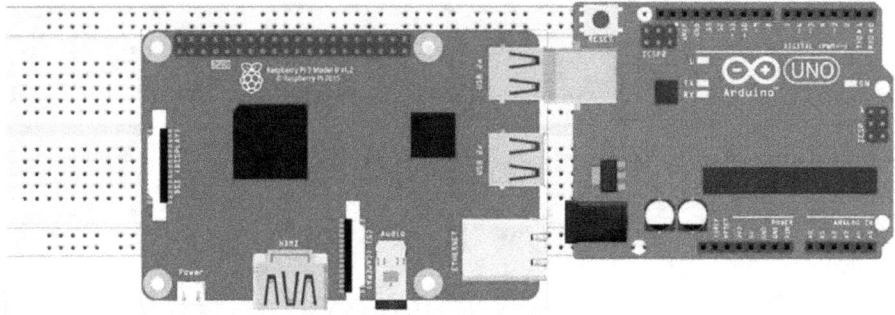

As you can see from this very simple image that the basic connection scheme is connecting this you ask to be born to this you must be bought. It's very simple. We already said that we will use this this cable. The one that comes with the Arduino board to connect these two boards together. Now let's see this and our simulation software as you can see here. Let's get Arduino OK. Now this is our board. And the reality all yours use that cable. But for now this is our USP and this is you can just bring them close to each other. OK. That's it. These three balls must be connected to each other using the cable that we mentioned earlier. Now it's very simple and basic connection.

FINISHING CONFIGURATION

```
sudo apt-get install python-serial

sudo pip install pyserial
```

Make sure the code is uploaded to arguido and your spirit by interface. Be sure to enable cereal and eye to see in body configuration. Go by menu then differences. Then tell us a bit about configuration. As you can see here then enable these two to see and serial communication then simply click OK. After this you need to start to rasberry by then open your terminal and execute these commands. The first command so to get installed by on Syria for serial communication. So they'll be installed by Syria for sterling by phone serial communication protocol. Now let's see key here is our window. Let's start the terminal OK. Now as you can see here we need to right so they'll get installed by phone see it via OK. Seems that there is a problem since I already was in told this because OK we need put on it. So. So don't be cagey for the packaging. OK suiting up by phone Sylvio. Now let's go ahead and enable the web. So the bib and so on by Syrian k it seems that we need to install Bebb OK. Now we need to install the configuration menu. But let's read the first sentence K so that I would get in a store and buy some cereal so pies and cereals are ready to burn.

```
pi@raspberry ~                                    _ □ x
File  Edit  Tabs  Help
/dev/tty1    /dev/tty20   /dev/tty31   /dev/tty42   /dev/tty53   /dev/tty7
/dev/tty10   /dev/tty21   /dev/tty32   /dev/tty43   /dev/tty54   /dev/tty8
/dev/tty11   /dev/tty22   /dev/tty33   /dev/tty44   /dev/tty55   /dev/tty9
/dev/tty12   /dev/tty23   /dev/tty34   /dev/tty45   /dev/tty56   /dev/ttyACM0
/dev/tty13   /dev/tty24   /dev/tty35   /dev/tty46   /dev/tty57   /dev/ttyAMA0
/dev/tty14   /dev/tty25   /dev/tty36   /dev/tty47   /dev/tty58   /dev/ttyprintk
/dev/tty15   /dev/tty26   /dev/tty37   /dev/tty48   /dev/tty59
/dev/tty16   /dev/tty27   /dev/tty38   /dev/tty49   /dev/tty6
/dev/tty17   /dev/tty28   /dev/tty39   /dev/tty5    /dev/tty60
/dev/tty18   /dev/tty29   /dev/tty4    /dev/tty50   /dev/tty61
pi@raspberry:~ $ ls /dev/tty*
/dev/tty     /dev/tty19   /dev/tty3    /dev/tty40   /dev/tty51   /dev/tty62
/dev/tty0    /dev/tty2    /dev/tty30   /dev/tty41   /dev/tty52   /dev/tty63
/dev/tty1    /dev/tty20   /dev/tty31   /dev/tty42   /dev/tty53   /dev/tty7
/dev/tty10   /dev/tty21   /dev/tty32   /dev/tty43   /dev/tty54   /dev/tty8
/dev/tty11   /dev/tty22   /dev/tty33   /dev/tty44   /dev/tty55   /dev/tty9
/dev/tty12   /dev/tty23   /dev/tty34   /dev/tty45   /dev/tty56   /dev/ttyACM0
/dev/tty13   /dev/tty24   /dev/tty35   /dev/tty46   /dev/tty57   /dev/ttyAMA0
/dev/tty14   /dev/tty25   /dev/tty36   /dev/tty47   /dev/tty58   /dev/ttyprintk
/dev/tty15   /dev/tty26   /dev/tty37   /dev/tty48   /dev/tty59
/dev/tty16   /dev/tty27   /dev/tty38   /dev/tty49   /dev/tty6
/dev/tty17   /dev/tty28   /dev/tty39   /dev/tty5    /dev/tty60
/dev/tty18   /dev/tty29   /dev/tty4    /dev/tty50   /dev/tty61
pi@raspberry:~ $
```

Now let's look to the command for installing bib. As you can see here we have different ensembles. We need to sort of this library but let's say we need to update the libraries at first so we get a date K as you can see it's updating SLUB please. After that we will use this sentence to install the bib that we need to install. That's the area of communication as you can see whenever you face a problem you must search for the solution. Don't just on this problem and finish the lesson or say I can do it you have to search engines to look for someone to help you look for onsides OK it's updating. You need fewer minutes on stuff than when you get back here. But let's see what else we need to do. And we need to connect what we know to our spirit by using the unspeakable and execute this order allow us or list devices DTI to list all or to get a list of USP devices or Sierre connection devices then find online with dev t t y a C zero or something like Dev. Why is he on one cheek for on SC I'm I don't know about Z in a one or two as you can see. We have lots and lots of devices. Here we have the one that is no other issue. As you can see so we have to edit our code. I'm one of the I'm 0 so open bison again and change that I see. Equals serial serial. T y se 0 to see someone 9600 to the SE number you've found so if in your case you got zero then the line should look say it equal to the serial Dev D t y se I'm zero 9600. But in our case here we got one so we need to keep it on one. This is a very important step that you must make so that you allow of communication between all of us by the by if you not 0 1 it will be narcism communication because once the Arduino board has been about we won't see the Arduino board

THE BREADBOARD

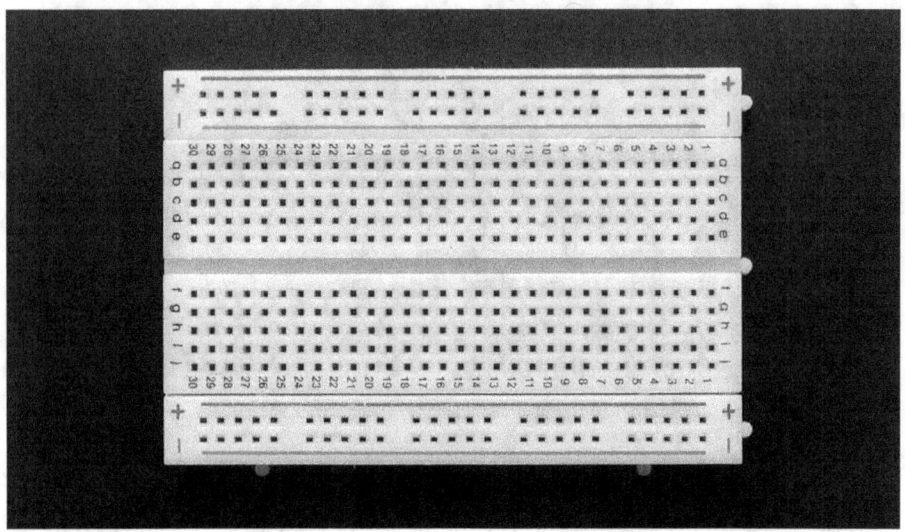

Now, this is a breadboard where you plug things to speaking of plugs here's you can see in the picture below you can use this type of customized wires or you can use regular wire and just strip the end. They are wires with little plug in bit belt on. You can just strip the insulation from regular copper wires. But these are easy because they come pre-cut and pre-stripped. Also they are collared so it's easy to tell them apart. Now that I've shown you those two things. Let me take a moment to expose the breadboard for you. Looks complicated with its letters and numbers but it's quite simple. The breadboard is just a short cut that's all.

You don't need it but it will allow you to test circuits without having to solder them together. Soldering if you don't know when you bind a specific type of metal around something to hold it together frequently used on the tips of wires the board creates specific electrical connections between its holes so that the wire don't have to be soldered. The only trick we need to know is where those connections are. But I will show you and explain to you everything. As you can see in this picture so I will explain each location on the board. There are four areas needed to be clarified the areas numbers as you can see. We have one area to area three and four for Area 1 and 2 other the blocks in area one are connected to each other. It all the blogs and NEDIA do are connected to that. Does it provide areas for you to control the flow of power. Which circuit is going to have at least one wire attached to it attached to it. To block an area on and one wire attached to block an area to think of them as sides of rhetoric that all the water flows from one and through that through the device to the other end without completed circuit. Nothing will happen and of all on one floor. For the record area one with the plus sign is known as cathode y that it too was a minus sign is known as a. You don't need to remember that. I was just letting you know. Area 3 here. Inside there were three R5 of Blunk that are all connected to each other. For example if you block a wire in it into the air block on the far left and another ran into the earplug on the far right those two the those two wires will have a connection anyway and you plug anywhere else on the board will not be connected to the wires in there.12 inside. Inside your theory the plug can each have or are connected to themselves and nowhere else. Area for here. If you connected these two haul with our wire. The connection will extend into the row across from it. Why there are on the left and right side of the board share the same numbers. They are not connected. If you want them to share a connection you need to bridge that gap with wire. Again this is this is just a short cut fall saw doing the wires to their respective location. If you want to incorporate Arduino or build some nobbled of some sort you absolutely don't want to use a bead board but instead you want to just Soledad all the parts together. However the bread bar is a great teaching tool.

WHAT IS A MICROCONTROLLER

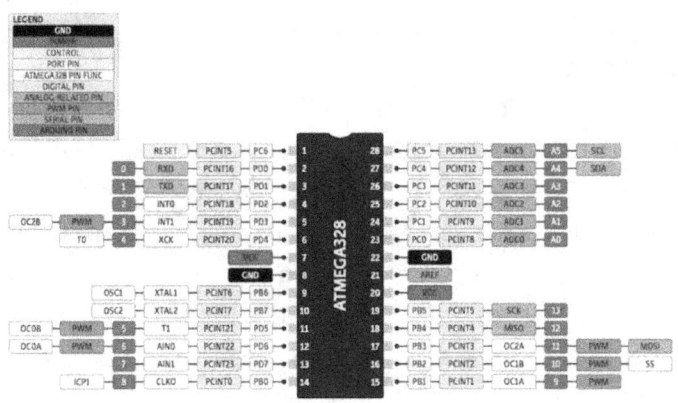

Micro wonder about the microcontroller. First of all, we need to understand the difference between a microcontroller and a microprocessor. Well if we say the microprocessor As the brain of a computer then you can say that microcontroller work as a brain as well as a muscle for the computer. In other words, the microcontroller is a system on a chip. It means that on the chip, it has its ram that is four kilobytes. It has its hard disk, there is 32 kilobyte, and it had its processor that is of eight megahertz. Well, these systems and specification sound very low, but they can run almost all of the required programs in thighs many only thighs small amount of memory. Well about microcontroller in its architecture. In three to eight, we have 28 pins. You can see here, thighs are the physical number of the pins and in the pink one, you can see that the number that is assigned in Harding, you can see that PIN four, four in the physical pin is PIN two in Arduino. In the whole section, we are going to deal with the Arduino the PINs on the Arduino. Okay, now you can see here that microcontroller is eight, what is it what will it be 10 set as the memory structure or the architecture of the Z there are different type of microcontroller which has a bit 16 bit 32 bit and 64-bit architecture, but here we are going to build with the eight-bit.

HOW DOES ARDUINO THINKS

```
int LED=13;
// the setup function runs once when you press reset or power the board
void setup() {
    // initialize digital pin LED as an output.
    pinMode(LED, OUTPUT);
}

// the loop function runs over and over again forever
void loop() {
    digitalWrite(LED, HIGH);   // turn the LED on (HIGH is the voltage level)
    delay(1000);               // wait for a second
    digitalWrite(LED, LOW);    // turn the LED off by making the voltage LOW
    delay(1000);               // wait for a second
}
```

The one in which we will talk about how does I do think so how does one think how do we know is it programmed with ones and zeroes. But hopefully that's not how we programmers have to write the code. Most are Gram's are born through the use of C++ and C to a very common computer languages. Now you may be wondering what the computer language even has a computer language is a tool used to develop a set of instruction. A computer is capable of understanding. In this case my computer is and I didn't hand the set of instructions are quite some in at least in this lesson. I looked into the code how very simple Arduino script usually starts out with an initialization or a variable to create an integer value of 1 and is used for example. This line of code creates an integer value of all calls and assigns the value of 13. Now we need we need to do something with with this allocated valuable. The next step is to create a function in which we will use the data stored in lead to turn on physical and voids it up.

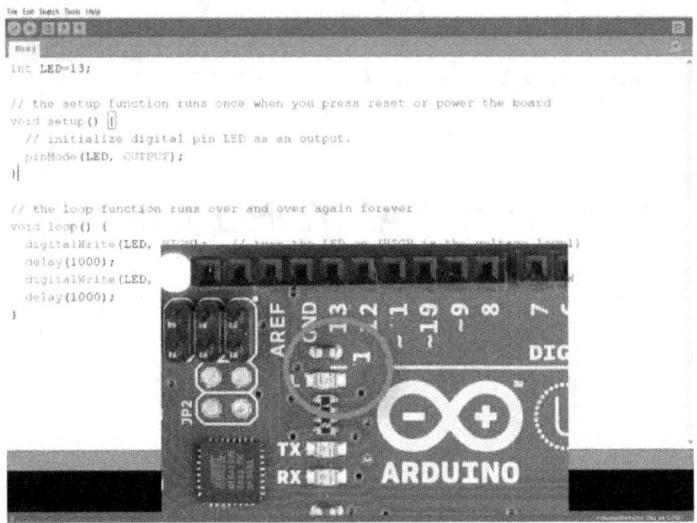

Basically this does is create a function from which no value is returned hence void the function in the context of this Arduino program gives you the opportunity to configure any physical been in or out what and which when to use this line as shown here and then mod lead out what this assigns the bin number to 13 and you start to think about what would have been mugged. Let's Arduino or what about to change the settings on a since it means 30 an hour. Do you mind we can simply what. And this also makes it much easier to change the ban in the future as we can as we can just change the variable to different integer density and Adreno happens to be an instant. So it's quite easy to try out the subprogram without any extra hardware. Apart from the device itself you can clearly see that circles in the image below or if you have hardware have one. You can't add you only like so there is still one but missing the actual instruction at the moment that we know knows where the digital.

Set Up

- Power the board by connecting it to a PC via USB cable
- Launch the Arduino IDE
- Set the board type and the port for the board
- TOOLS -> BOARD -> select your board
- TOOLS -> PORT -> select your port

The basic setup is the power from the board is derived. As you can see, the power from the board is derived from the PC using the USB. if you're initially for testing purposes, you will be connecting it to the PC to upload your program. And when you're running it in standalone mode when your program has been uploaded on this board, you can run it from this

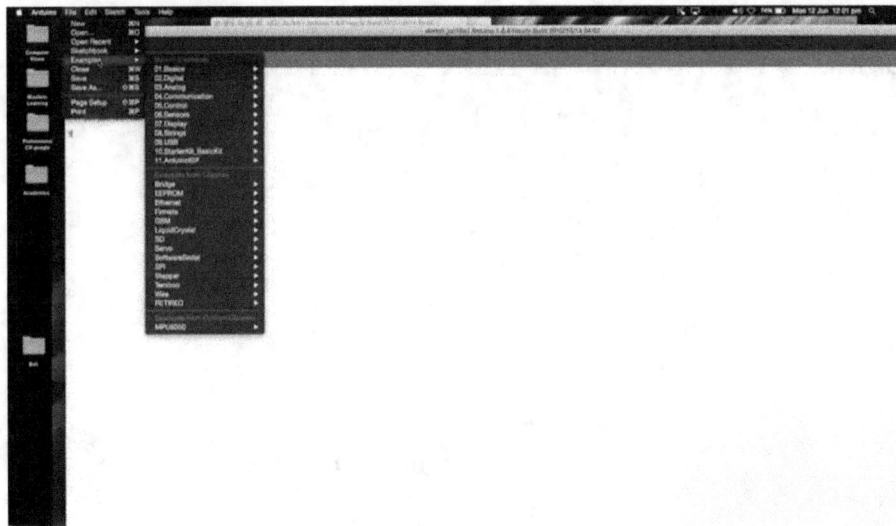

power supply input. You plug in a DC adapter of five volts and it's going to work fine. Then next to the last ID then select the board type. let me show you. over here I have my Arduino ID as you can see you have me basic functionalities. This is the code verification button. This is the code upload button. Then you have the File menu, you can create a new sketch. A sketch is a program you write for an Arduino. Then you can open a new, open an existing program, open recent programs and on. You even have examples, basic examples provided with the ID, which can work with various Arduino based boards. Now moving on to this sketch, the most important part is this tool whenever you connect your Arduino board I've connected my Arduino board to my PC now

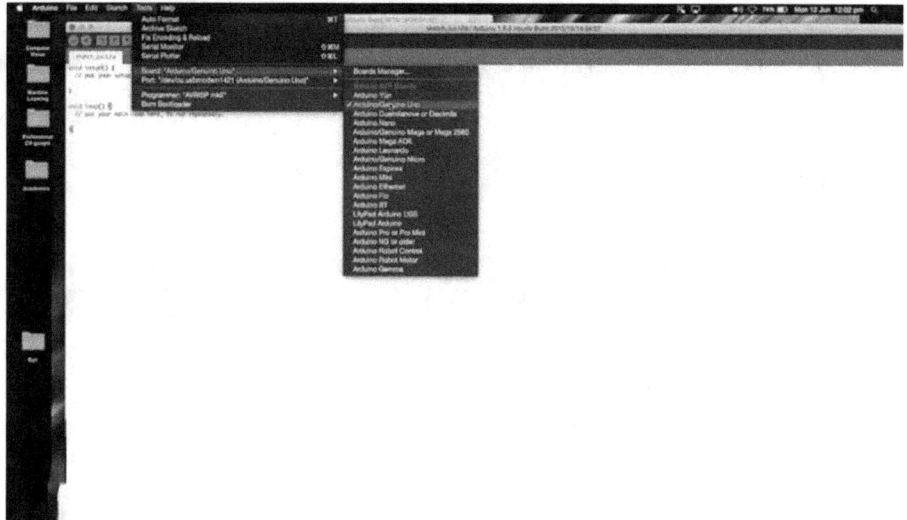

This tool these are the available boards. You can see since I'm using an Arduino Uno it has been automatically selected but in case it is not automatically selected you can choose the appropriate board

Arduino Uno. For my Mac, the port is already this USB modem 1421 you can see Arduino Uno has already been selected. now we are all set. Okay one more important thing is over here you can see this button this is the serial monitor. This is one of the good features of Arduino that while executing serial programs, you don't need to have an external conle or that kind of ftware, you can just use the inbuilt serial monitor to view the program. now once the board and the ports have been appropriately selected,

Set up (contd..)

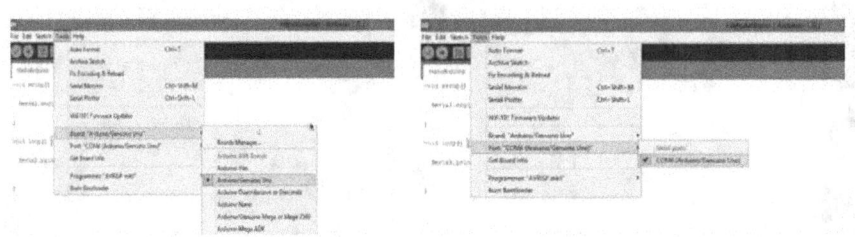

as I have told you, you select the board then the color Responding port in your PC for Windows based systems, it will be more or less direct it will show you a combat sport. It may be coming forward 1015 anything you choose it appropriately.

Arduino IDE Overview

Program coded in Arduino IDE is called a SKETCH

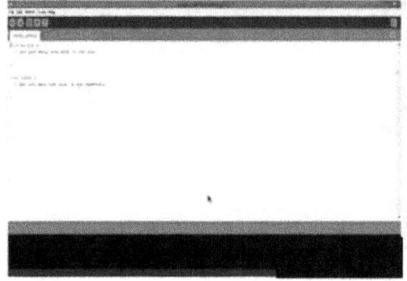

Then I've already shown you this. your Arduino sketch. As you remember the program which is written into the Arduino, for written for the Arduino is called a sketch. it consists mainly of two parts one is the setup, and one is called a loop. setup is analogous to for normal main, C or c++ based programs. The main function you use it is analogous to the setup

Arduino IDE Overview (contd..)

- To create a new sketch
  - File -> New
- To open an existing sketch
  - File -> open ->
- There are some basic ready-to-use sketches available in the EXAMPLES section
- File -> Examples -> select any program

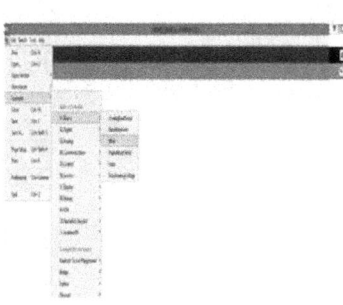

function in Arduino. And, as the name suggests, the loop function it is used for iteratively looping over instances. it's more or less common from file you click on new

Arduino IDE Overview (contd..)

- Verify: Checks the code for compilation errors
- Upload: Uploads the final code to the controller board
- New: Creates a new blank sketch with basic structure
- Open: Opens an existing sketch
- Save: Saves the current sketch

It will open a new file and you can try out various examples and sketches. we have al covered this one, this is the Verify button. the main feature is prior to uploading our code, if you have syntax errors or any such logical errors, it will be caught during verification it will say your compilation has failed. Once you pass this verification check, you can upload your code.

Arduino IDE Overview (contd..)

- Serial Monitor: Opens the serial console
- All the data printed to the console are displayed here

okay. Now, this we talked about this is the serial monitor, whatever data is transmitted through the serial port is printed on the serial monitor.

Sketch Structure

- A sketch can be divided into two parts:
    - Setup ()
    - Loop()
- The function setup() is the point where the code starts, just like the main() function in C and C++
- I/O Variables, pin modes are initialized in the Setup() function
- Loop() function, as the name suggests, iterates the specified task in the program

a sketch structure, as I've told you, it consists of two parts a set apart and the loop part. The function setup is the point where the Arduino compiler actually starts the code. it's just like animal West with the main function in C and c++ and various input output variables in modes whether you need to as you remember you have 14 digital input output pins. you have to explicitly tell your system whether you want to use the pin in a read mode or input mode or the output mode. Then the loop function is used for iteration. in this example code you can see we just use a serial port the inbuilt serial port, we just write serial dot begin 90 690 600 is the baud rate you can have various baud rates. We'll come to that we'll cover that in the consecutive lectures. You can have various borders for different systems but 9600 is more or less the commonly used baud rate for most of the systems. And within void loop you want to iteratively loop this Hello Arduino. this serial dot print ln that If you write serial dot print, it just prints the Hello Arduino string. Otherwise if you write print ln ln is actually a new line. it will print hello Arduino in a new line.

Supported Datatype

- Arduino supports the following data types-

Void	Long
Int	Char
Boolean	Unsigned char
Byte	Unsigned int
Word	Unsigned long
Float	Double
Array	String-char array
String-object	Short

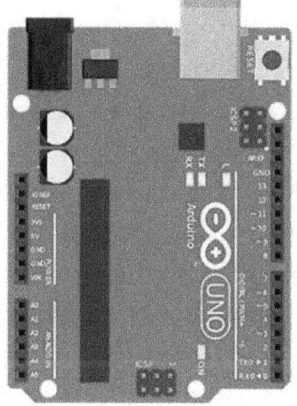

prior to this will look at the sample code. as you can see in this Hello Arduino code within the word setup, we have written serial dot begin 9600 and within void loop, we have just written serial dot print ln Hello Arduino.

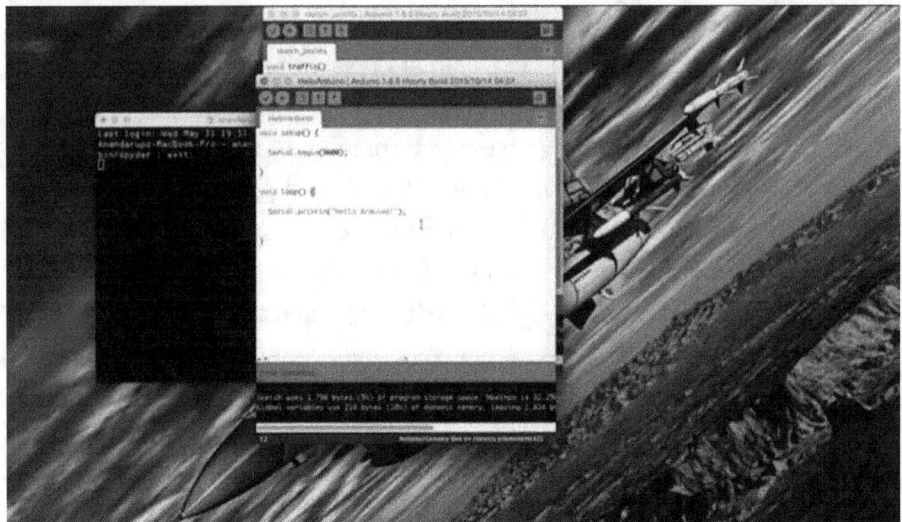

Now prior to doing anything Verify the code. As you can see, it is compiling the sketch. If your code compilation is correct, it's correct. It just shows how much memory it is using and all those. If if you're in error suppose I delete this semicolon now again I verify the sketch it will give you an error. this is a good practice prior to uploading blindly you just verify your code. Okay, now the code verification is successful. The ports have already been chosen, I upload my code. it is now compiling the sketch and it is uploading it to the Arduino board. Now, the code has been uploaded to the Arduino board since The function of this program is to print hello Arduino on the serial port iteratively will open the serial monitor. As you can see, it is printing Hello Arduino. Right? it is actually quite fast, we can actually modify it will put up a function called delay,

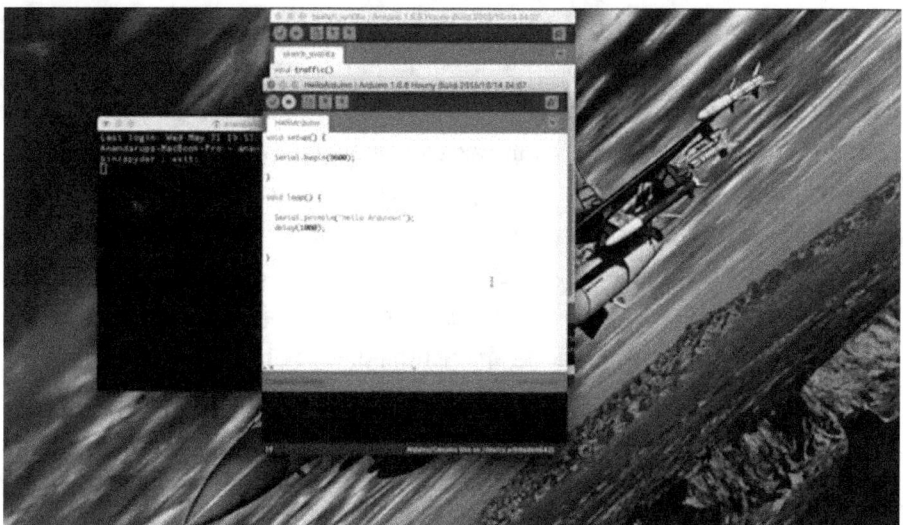

let's say a delay of one second. this thousand is actually the delay in milliseconds we'll put up a delay of one second. The code has been verified we upload it again. Now again we open the serial monitor. Now can see now the delay has been increased it prints after one second I hope this was easy. Now let's move on to the next menu. Okay, like other programs, Arduino al supports various data types you have void in Boolean byte word, float array, string object, long cat unsigned cat. It is mewhat similar to your normal c programs.

Arduino Function Libraries

- Input/Output Functions:
    - The arduino pins can be configured to act as input or output pins using the pinMode() function

            Void setup ()
            {
            pinMode (pin , mode);
            }
    Pin- pin number on the Arduino board
    Mode- INPUT/OUTPUT

Arduino has lots and lots of libraries. Since it's an open urce platform.

collaboratively people al people and companies and organizations, they upload their own Arduino libraries. for most of the functions will obviously get easy access to various libraries. as you already know, the pins can we configure to act as input or output depending on your requirement. to do this, this function pin mode is used. you can see the syntaxes pin mode, pin, comma mode. this pin is the number, the pin number actual pin number on the Arduino board as you can, if you focus on this board, you will see various pin numbers I have written over here 123 since these are the digital pins, it is al written digital. in the pin mode against the pin, you just write the number of the pin.

it's that simple and the mode you just write input or output if you want the pins to work in input mode, like you're connecting various senrs to it, which you will acquire senry inputs. You put the pin in input mode and if you want to activate mething, maybe a light on Led or a motor, you put the pin in output mode.

Arduino Function Libraries (contd..)

- digitalWrite() : Writes a HIGH or LOW value to a digital pin

- analogRead() : Reads from the analog input pin i.e., voltage applied across the pin

- Character functions such as isdigit(), isalpha(), isalnum(), isxdigit(), islower(), isupper(), isspace() return 1(true) or 0(false)

- Delay() function is one of the most common time manipulation function used to provide a delay of specified time. It accepts integer value (time in miliseconds)

ARDUINO SOFTWARE FOR THE FIRST TIMERS

The next module Arduino software for the first time. In the thighs module, we are going to discuss Arduino and what is a red board. Let's get started with programming download and install of the software in about the Integrated Development Environment and writing our first program that is blinking an LED or you can say hello world. Before going there. I want you to go to the chart to go to Google and search for Arduino.

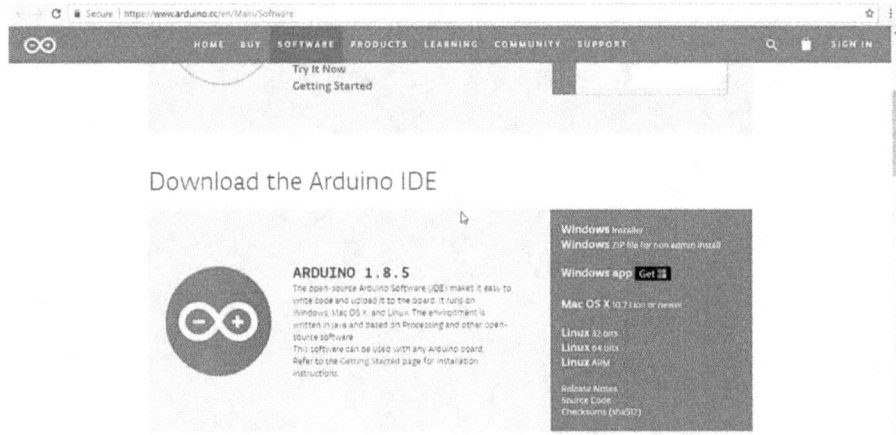

After searching Arduino, you can click on Arduino button Arduino home and after doing that, just click on the software after clicking on the software you have to select your type of installation. It could be a window It could be a Mac or it could be a Linux for me I have selected Windows Installer after clicking on that you will get to the next page that is telling about contribute to download or just for our case right now we are going to just download after the When the download of the Arduino software will be in well I have downloaded it earlier. So you can just see here how the installation of the Artemis okay. Thighs are Arduino 1.8 point four that is for Windows.

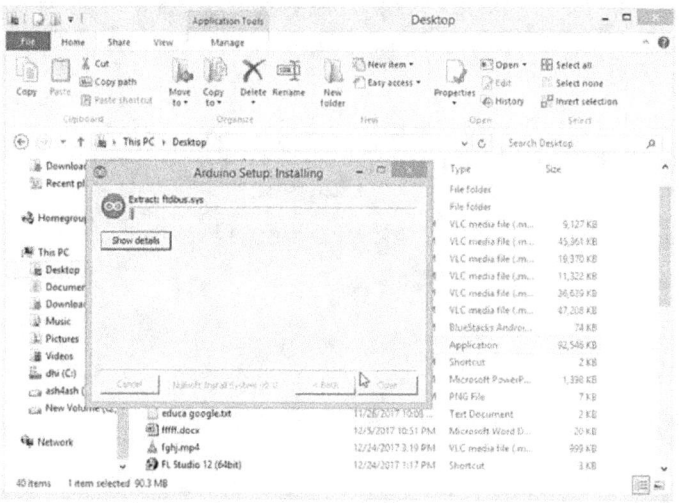

So it is an executable file. Just click on it and it will start to ask the security question just click yes on it. Click on I agree. Click on Next click on the location it will ask for the location where the agenda should be installed. For me, it sees program policies and the folder name Arduino. After clicking it, it will just begin to install. If you want to see how the things and how the libraries or files are going to stall, just click on Good as you can see here, well within a minute it will get installed. Well, I have just it will ask for the driver that should be installed in the Arduino. So just closing it. Okay. Now on the desktop, you can see that there is an icon, which is named Arduino. Okay, just see there. If you don't find the icon, don't worry. Go to C. Go to Program Files x86, click on Arduino and there you can find the icon that is Arduino e xe just create a shortcut or have it on the textile and you can begin Okay, click on it will ask for the access for the Java file click on Yes. And now you can see the rd. Thighs are the integrated development environment for Arduino here you can see that it has all the features that are needed for us for the board if you can garden of water you can find here the port and all the necessary buttons.

Content

- Operators in Arduino
- Control Statement
- Loops
- Arrays
- String
- Math Library
- Random Number
- Interrupts
- Example Program

we'll cover the basic topics, the operators in Arduino, control statements, loops, arrays, strings, the mathematics library, random number interrupts and our example program which will be a bit complicated than the previous one.

Operators

- Arithmetic Operators: =, +, -, *, /, %
- Comparison Operator: ==, !=, <, >, <=, >=
- Boolean Operator: &&, ||, !
- Bitwise Operator: &, |, ^, ~, <<, >>
- Compound Operator: ++, --, +=, -=, *=, /=, %=, |=, &=

So, basic operators as normal c, c++ or Python programming or other languages you have the basic equal to plus minus multiplication division modular division operators then comparison operators If you have equal to not equal to less than, greater than, and all those operators then you have Boolean operators bitwise operators and compound operators.

Control Statement

- If statement
  - if(condition){
        Statements if the
        condition is true ;
        }
- If...Else statement
  - if(condition){
        Statements if the
        condition is true;
        }
        else{
        Statements if the
        condition is false;
        }

- If.......Elseif.....Else
  - if (condition1){
        Statements if the
        condition1 is true;
        }
        else if (condition2){
        Statements if the
        condition1 is false
        and condition2 is true;
        }
        else{
        Statements if both the
        conditions are false;
        }

Moving on to control statements, these will basically cover the various checking and looping parts. So, a normal FL statement in Arduino, you start off with the if statement. So, if you have a condition and within these curly braces, if the statement condition is true, if else if another statement condition is true, or else if none of the above statements are true, then this loop will execute. Moving on to switch case, you have switch and Joyce's have case option one and statement and then a break function.

Control Statement (contd..)

- Switch Case
  - Switch(choice)
        {
        case opt1: statement_1;break;
        case opt2: statement_2;break;
        case opt3: statement_3;break;

        case default: statement_default; break;
        }

- Conditional Operator.
  - Val=(condition)?(Statement1): (Statement2)

To each case, so, case option to your statement to then again a break and so on. And at the end you have a default case after that, you again have a break function, then you have a conditional operator will avoid using conditional operators such as these in Arduino. So, it is condition if it is true, it will execute statement one else it will execute Statement two these kinds of statement operators are best avoided during Arduino programming.

Loops

- For loop
    - for(initialization; condition; increment){
        Statement till the condition is true;
        }
- While loop
    - while(condition){
        Statement till the condition is true;
        }
- Do... While loop
    - do{
        Statement till the condition is true;
        }while(condition);

So, in loops, you have the basic for loop then you have the while loop, you have a do while loop. These are pretty common example you have a nested loop that is a loop inside another loop you can have many nested loops inside each other. So, they will have an infinite loop

Loops (contd..)

- Nested loop: Calling a loop inside another loop

- Infinite loop: Condition of the loop is always true, the loop will never terminate

So, to run an infinite loop for example, you all you need you develop a system in which in which you need to turn on and off or light or LED or any other device infinitely as long as the device is on your system is checking. So, recall from the last lecture, which I showed the blinking LED example. So, you can see, if we put it inside an infinite loop, as long as the Arduino board is powered, it will keep on blinking. So, your functions can be made more complicated. Instead of LED you can have motors instead of motors, you can have actually cameras mounted on the motors and they keep on rotating you have you can have a multitude of sensors which are interfaced with the cameras and the motors. So, you can for example, you can build a security system which will keep on running as long as your processor works. is fine and the power is being supplied you can always connect it to a battery supply to generate power for it.

Arrays

- Collection of elements having homogenous datatype that are stored in adjacent memory location.
- The conventional starting index is 0.
- Declaration of array:
    <Datatype> array_name[size];
    Ex: int arre[5];

Then you have arrays, arrays are a collection of elements having homogeneous data type and which are stored in and just in memory locations. The conventional starting index is zero in Arduino. So, declaration of array you just start off with a data type it may be an array of integers, so int array name and the size. So, for example, in an array, this is a variable name array five it will allocate five spaces for your array.

Arrays (contd..)

- Alternative Declaration:
  int arre[]={0,1,2,3,4};
  int arre[5]={0,1,2};
- Multi-dimentional array Declaration:
  <Datatype> array_name[n1] [n2][n3]....;
  Ex: int arre[row][col][height];

Then you can have alternate declarations. Suppose int array and this blank bracket equal to within these curly braces you have 01234 So, these will be automatically In this area, then again you have entered a five you can just put in three variable three values inside this array and the remaining will be kept blank, maybe for later use, you can fill those also when you have multi dimensional array declarations same as the previous one, you have the data type the array name, then the dimensions for the first dimension Let it be n one n two n three. So, for example, if you want to declare an array for image which normal normal RGB image, so, you have three channels red, green and blue. So, each image will have a 2d structure with rows and columns. And there will be a depth for each RG and b. So, maybe for those types of data, you have int array row column

String

- Array of characters with NULL as termination is termed as a String.
- Declaration using Array:
  - char str[]="ABCD";
  - char str[4];
    - str[0]='A';
    - str[0]='B';
    - str[0]='C';
    - str[0]=0;
- Declaration using String Object:
  - String str="ABC";

Then moving on to string string is an array of characters with null as the termination declaration is maybe using Cat Cat string. Here str is the array. So ABCD so this is stored in str care str four and you can individually access each IDs you can store a b c or maybe zero. So this is using the same location if you want to individually stored in different locations. So sorry come in the same location. If you keep on storing this, the last one will be last character stored will be updated, other will be overwritten. If you want to store in different locations you just change it from string zero str one str. str three And so on. So you'll have consecutive ABC zero, side by side in these locations. Another thing you can also have you also have a data type string. So string str equal to ABC will give you ABC altogether you don't have to store in individual locations. So this is one of the benefits of using Arduino.

String (contd..)

- Functions of String Object:
    - str.ToUpperCase(): change all the characters of str to upper case
    - str.replace(str1,str2): is str1 is the sub string of str then it will be replaced by str2
    - str.length(): returns the length of the string without considering null

So some commonly used functions of string. So str to uppercase point to notice to uppercase, t, u and c are caps. So this has to be followed strictly since this is part of the syntax, so it changes all the characters of string to uppercase. Then you have string str dot replace string one and string two. So, string one,

Math Library

- To apply the math functions and mathematical constants, "**MATH.h**" header files is needed to be included.
- Functions:
    - cos(double radian);
    - sin(double radian);
    - tan(double radian);
    - fabs(double val);
    - fmod(double val1, double val2);

if it is substring of str then it will be replaced by str n str dot length it returns the length of the string without considering the null character. Then another commonly used library is the math library to apply the math functions, the math dot h header must be initially called otherwise, you won't be able to access these functions. So, some of the common functions are cos, which is in double radians in sine, tan, f floating absolute fabs, right floating mod. So double value one and double value two. So you have two values and F mod will give you the modular division and the result will be a floating point number. Then, continuing with the mat library again, you have XP which signify six By the exponential function you have log function, this will give you the natural logarithm of the value then you have log 10 and you have square function power function. first argument is the base, the second argument signifies the power.

Random Number

- randomSeed(int v): reset the pseudo-random number generator with seed value v
- random(maxi)=gives a random number within the range [0,maxi]
- random(mini,maxi)=gives a random number within the range [mini,maxi]

Then another commonly used example is random number. So, one of the functions of this random number is random seed. So, the syntax is random seed as capital you need to focus on this one because this is the inbuilt syntax for Arduino. So, random seed int V, it resets the pseudo random number generator with seed value B. So, you already the seed value is the starting point from which the random number will initialize its function So, you give a starting value from it, the random number will generate, then random macside, it gives a random number within the range zero to max height, then you have random mini and Maxi, it gives a random number within the range, mini and Max.

Interrupts

- An external signal for which system blocks the current running process to process that signal
- Types:
    - Hardware interrupt
    - Software interrupt
- digitalPinToInterrupt(pin): Change actual digital pin to the specific interrupt number.
- attachInterrupt(digitalPinToInterrupt(pin), ISR, mode);
    - ISR: a interrupt service routine have to be defined

Then moving on to interrupts you have an external signal interrupts these are basically an external signal for which the system blocks the current running process till receiving that signal. So, basically, you have two types of interrupts one is hardware and another is software. So, I'll give you an example. Suppose, you're in a loop, you're waiting for a checking condition, whether that checking condition holds true or not. And maybe from an external source. You're getting that check In condition, for example, you have a button or a digital switch connected to our Arduino board. So, whenever you're pressing that switch, your system will blink an LED Otherwise, it will keep the LED off. So, this may be considered as a partially considered as an interrupt. So, this will be an external interrupt. So, as you can see digital pin to interrupt and then the pin number it actually changes that digital pin to the specific interrupt number, then attach interrupt digital pin to interrupt then pin then ASR then mode. So, is our is basically known as an interrupt service routine, it has to be defined explicitly. So, these are some of the more complicated functions so, we'll not focus on these

ARDUINO SYNTAX

Functions

- loop() and setup() are procedures
- You can create you own functions

Anatomy of a C++ function

Datatype of data returned,
any C datatype.
"void" if nothing is returned.

Parameters passed to function, any C datatype.

Function name

```
int myMultiplyFunction(int x, int y){
    int result;
    result = x * y;
    return result;
}
```

Return statement, datatype matches declaration.

Curly braces required.

```
void setup() {
}

void loop() {
}
```

Both setup() and loop() have no parameters and return no values

---

Syntax well syntax you can say that the syntax in a programming language is similar to grammar in English or any other language. So, the syntax is the method by which the program should be assembled properly. So that the compiler will come the compiler of the computer can read, you can say a compiler is a machine that is used to understand and check the program if it is right written in the syntax and then convert the C language the language that is simple to understand wires to hexadecimal language and hexadecimal language is consists of only two parts, that is 1010. Then after it can be written into each line can be written in hexadecimal code that contains a base of 16 into 123456789 and zero then a B, C, D, E, F, Thighs is the maximum strength of the hexadecimal value that is being written in the next time Okay, about understanding syntax, the Arduino is written in c++ language. So we don't need to have the basic knowledge for thighs course because we are doing it in practice and can getting we are getting a better understanding of the language if you have a prior-prior knowledge to that you can do it and it will be better for us for now if you have proper prior knowledge to it, it will be good for you because you will be better in programming and all other stuff. Okay, declaring variables and declaring prints well I have told you that a variable can be stated as an object or an instrument in which the values are put. For example, you can say that the x is equal to five it means a variable of x store the value five, you can change in within milliseconds, you can say the value of x can be changed US Census, one, eight or whatever. Thighs are known as variables, well, what about when I have told you in the Arduino there are about 28 pins out of which there are 30 digital inputs, five analogue pins that can be converted into details. So, we have about 10 digital pins and that can be working on the argument. So, we have to declare the pin with a variable. An important part is that a variable can also store the value of a pin, like if you had set a variable as is that I will be showing you an example that will be written. For example, if you have written the variable x is equal to 13 in you put all the places or x where the 13 has been returned, you can see that the program will be still working because the when the system goes to x, it will be taking the digital print the value of x is 13. So it will understand that x is equal to PIN 13. And when we put x in the place of Number as a variable, it will also understand

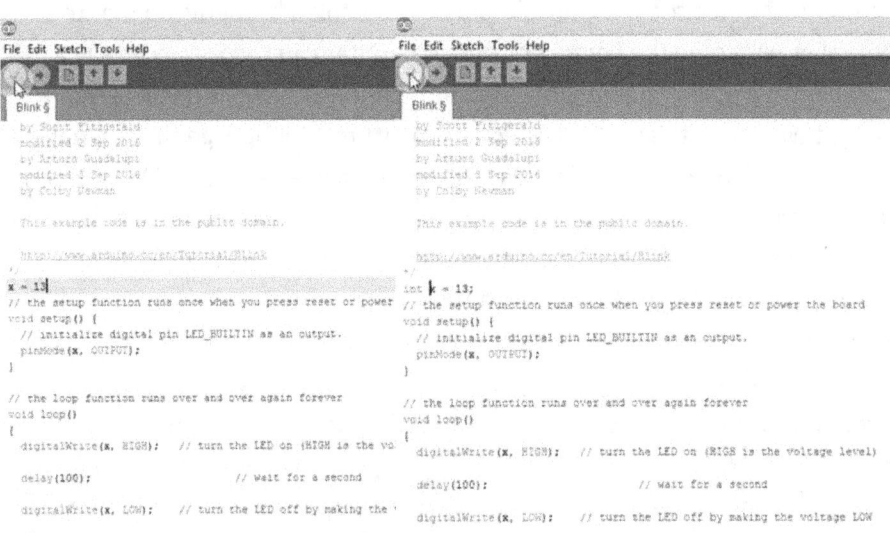

So, let us take it, first of all, we have to compile it thighs error you can see that error because we do not we had not to terminate, there is an error in the line because we have not terminated. So, let's terminated with a semicolon. Now, please take it as compiled version, you can say that the value of x is not declared and thighs are done by me to tell you that you cannot declare a variable directly without giving it a function without defining a function in because it is not Python we have to declare every variable with a function into where it can be afloat, it can be long or it can be a character I will be telling you about that in the next sections. But here we have to put it a value in t Okay, when we had put the value identity you can see that the colour has changed to blue it means that it is a village index. Okay, now, check it out. There are no compiling errors it means that the value of x is stored as a variable as value 13 and it has been stored and every point okay, let's just upload it to the world. First of all Diedre connections board, comport seven board, Arduino Uno. Now doing that upload, you can see that it's going and then uploading.

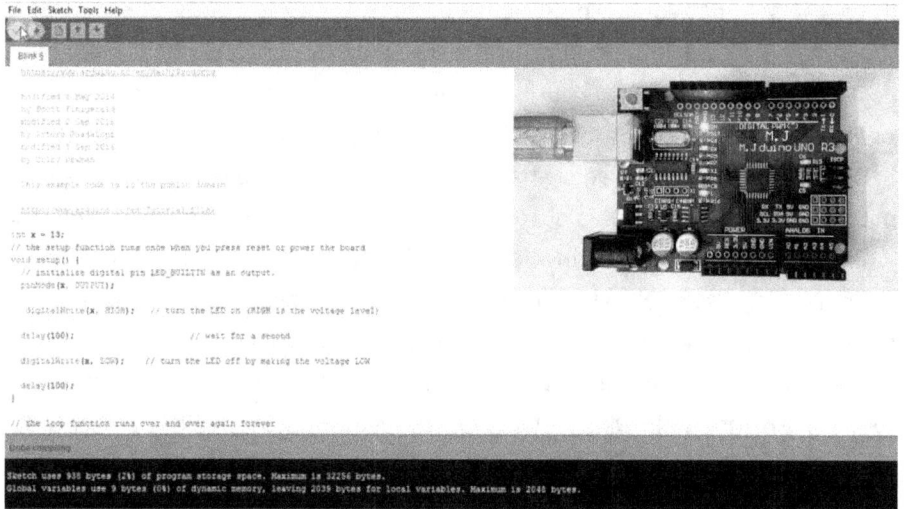

Now you can see that the board it's blinking it means that the method is correct. Okay, now let's get to the next part. After that, you can see that the next part is running one program at a time. While the difference we have to, we have to know the difference between running a program at one time and running it in a continuous loop. You can see there the blinking of LED is still continuous, it means that it is running in a loop. But what would happen if we run it is one time let's just take it We have to put it into the F we have to put it into Word setup, okay just a second let me do it come here to cut all the programs Ctrl x in write it here Ctrl V, you can see that after defining the been a program has been written with the same syntax Let us compile it okay after compiling it we can see that the program is written I am taking the microcontroller here so that you can check that it has been run for one time or not. I had completed the program has been running and you can see that it will blink for only one thing. Thighs are the method for the running program at one-time thighs is called void setup. And if you want to run the program for infinite long you have to put it in one room. Well, you can keep the word look empty but it is not possible to Key word setup empty because the declaration of a pin in the calling of the serial monitor and all the functions that are done here are placed in Word setup. Okay now let's get to the next part of it. Here you can see that running a program for multiple times or for running it in a long time. Okay, let's get back to the program, our ID we can see that your ID does go here, select the syntax program in curtir you can go to Boileau right in between the two curly brackets because if we do not write it between the curly bracket, it will say that it's an error okay. Again check ID integer is declared in the desert, our value our variable has been declared the world setup has been fulfilled with the condition when mod x is equal to 13. That is output Okay, now, let's go Get down to the word loop you can see that in Word loop digital right x is too high for 100 milliseconds, x is equal to low for next hundred milliseconds it is written in the loop so it will write it will run infinite longer for infinite room okay let's check it and upload of the program is done and you can see that it is running for infinitely long time Okay, let's get back to another section okay after that we will come to see what is a function if we run a program for one time it is ok but if we are going to call the values for multiple times, we need a function in which the program is returned for one time and whenever we want to call it for multiple times we will call the function I just give you a simple example look here.

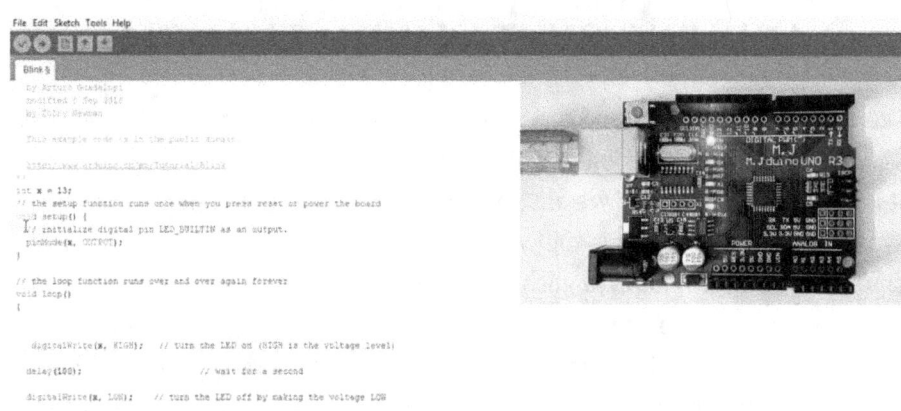

Okay here cut all the program with control x. Okay, after would set up before Lu v right v o ID void glow g l o w you can see that void low has been written give you two curly brackets, after the given two brackets after that give it two curly brackets, okay. In between them, paste the program and in the word loop called the functions g, l or w give them brackets in terminated with the semicolon, you can see that the function has been called here and its value has been returned here. Now, let's just take it we have checked it, there is no comply compilation error. And we have uploaded it to the board and it's successfully uploaded and you can see there the blink has been still going on. Okay, the function method is correct.

ARDUINO SYNTAX WITH EXAMPLE

We are going to understand the basics of programming or I meant to say the syntax that is used in Arduino programming. Okay and let's begin the lecture. It is in Module Three and we are going to start it. Okay, let's see about the syntax, the syntax as I have told you earlier that the syntax of the program is similar to the grammar in English. Okay, let us see the key points of it. First of all, here we can see a semicolon the semicolon in the programming is defined as a terminating statement. For example, if we select an integer and define the value to a variable, let's just say we have defined a value integer x is equal to 13.

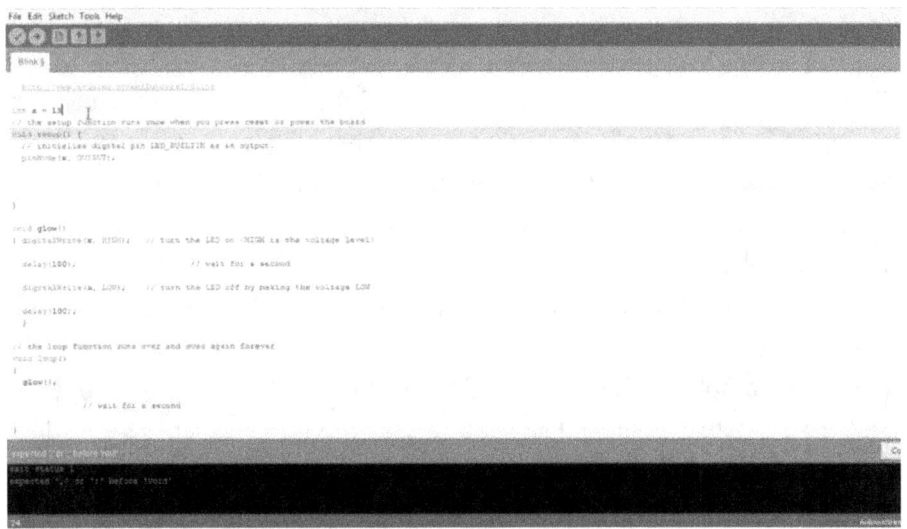

But if we try to compile it, after compiling, you can see that there is an error. There are a simple eight need a comma or a semicolon after the statement so that it can terminate okay. Let's just put a semicolon or a date. Now see that it's perfect, but something that is to remember that always after calling a function, there should be no semicolon.

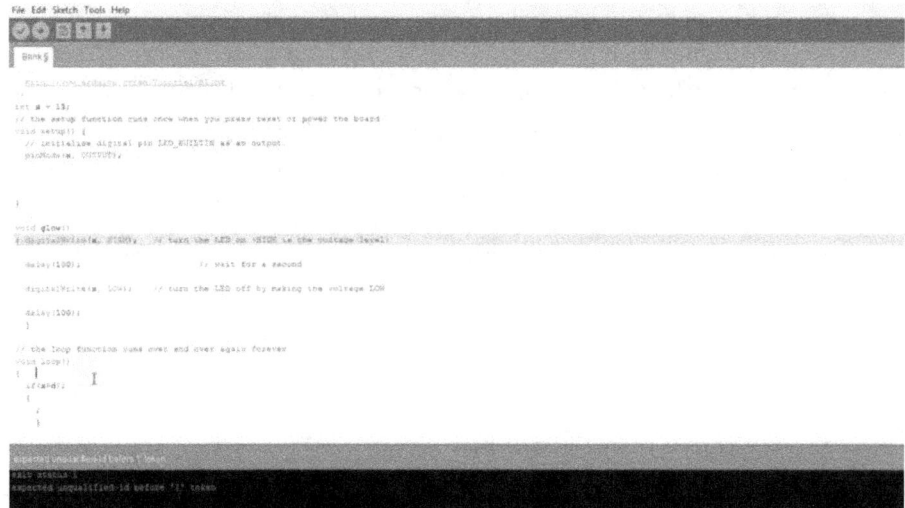

If you put a semicolon after a function declared, it will be said that it is not possible or it will show it will give an error. Okay. Now another next thing is that if we call something a statement such as if statement, for example, say if, let's just say if x is greater than five, remember something that after calling a statement, there should be no semicolon the statement will be terminated with the help of two curly braces, there should be no semicolon that should be placed in it.

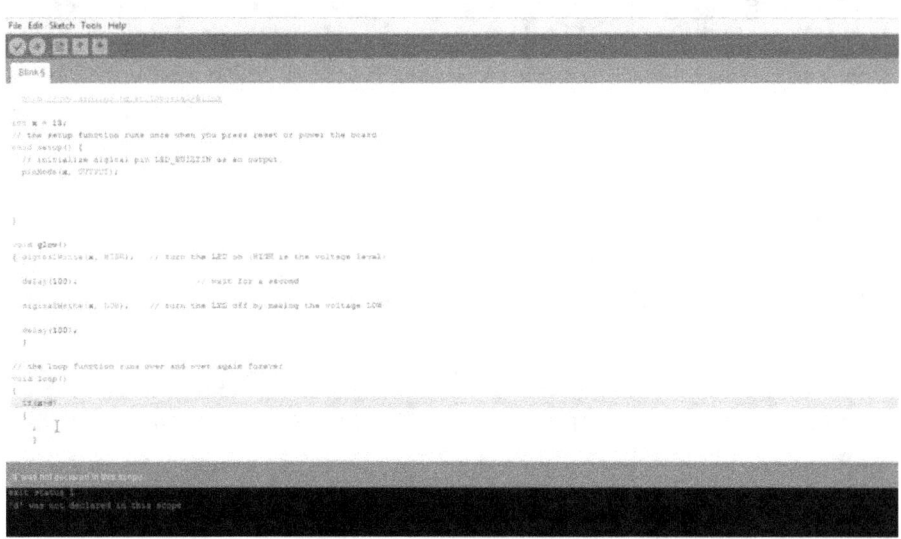

And another important thing there the semicolon inside another important function is dead. Our founder's curly braces there should be a simple semicolon is placed or else it will give us an error. First of all, let us see the correct version of it. It is giving us an error, the error is invalid low, D was not declared. Okay, let's just clear it to give us five. Okay, let's just compile it. it compiles perfectly just delete the semicolon after that. It will give us an error okay it is not.

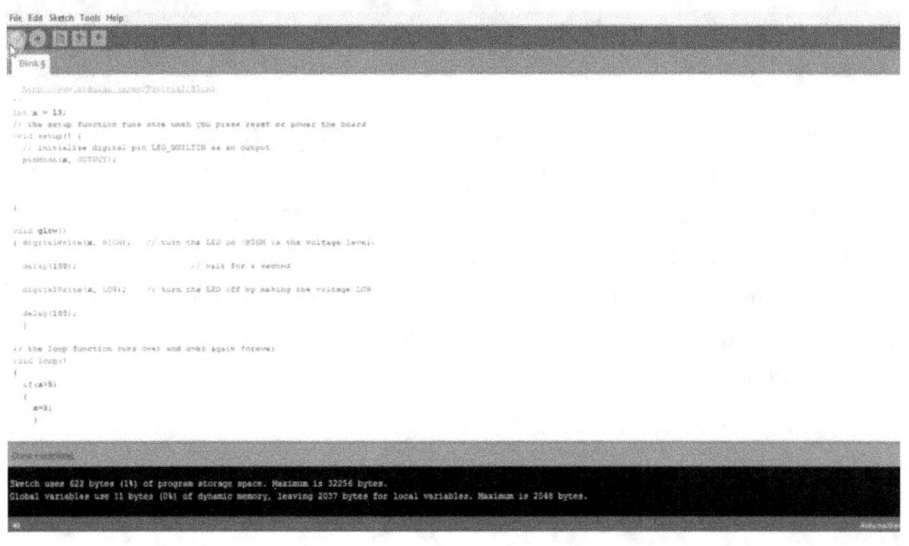

You can say that you can make an empty statement out of that it is not necessary to put a semicolon but it is for a better practice that after the statement is given, such as x is equal to three, you should terminate it with a semicolon. Let's just check it. Well, the semicolon is in place. The termination of the statement with a semicolon comes makes it compile level for the compiler and it can be converted into hexadecimal values that are fed into To the cows that are fed into the micro antelope or Okay, let's just sit another part of it. After that, we can see that there are two backslashes. What are they? Let's see backslashes in rdno are used to comment a line.

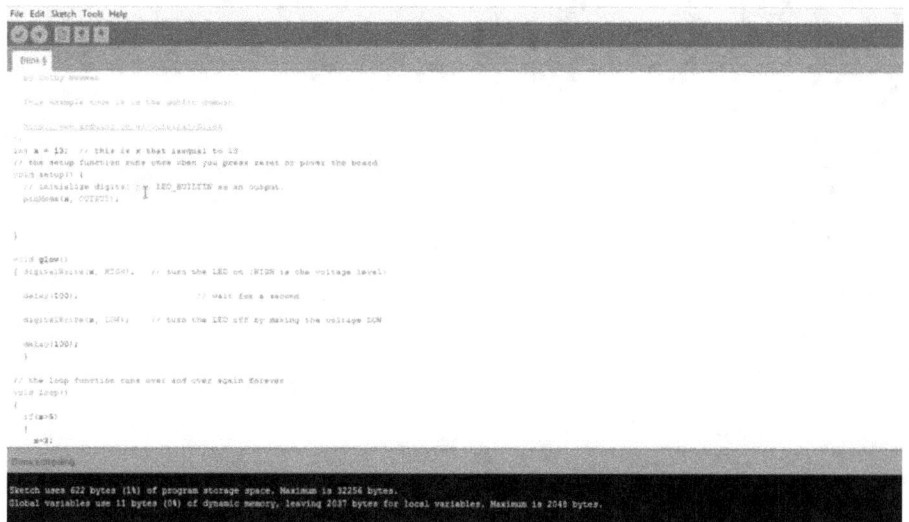

For comment, I want to tell you that if I make something like x is equal to five, no problem x is given is equal to a value for just do a correction, let me do a correction. I'm going to the first line See, you can see that integer x is equal to 13. But if you want to tell some another programmer or if you want to tell the future of yourself that what is your return 10 you can write a simple line in English without making it go inside the compiler. Give two semicolons after that, give to do two backslashes after that and write something that thighs are highs exit that is equal to 13 the line will not go into compiler legislated.

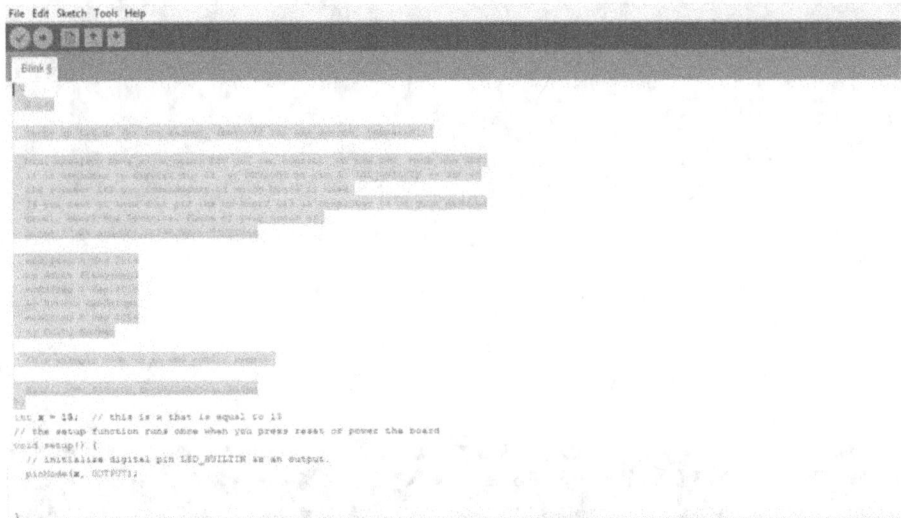

The line is perfect, but what happens if you want to compile multiple numbers of lines here you can see that there are multiple numbers of liner compiled. So let's just practice it. Thighs are done with the function backslash n n Astrix for the beginning and something important that should be remembered that a backslash then now for termination, it should be done by a backslash for the termination of the line, it should be done with the help of an Asterix and a backslash. Symbols index, backslash and Astrix for studying sticks and blacks backslash for a termination right something.

```
File Edit Sketch Tools Help

Blink §

/*
 * hello
 * world
 * this is a blink
 */
int x = 13;  // this is a that is equal to 13
// the setup function runs once when you press reset or power the board
void setup() {
  // initialise digital pin LED_BUILTIN as an output
  pinMode(x, OUTPUT);
}

void glow()
{ digitalWrite(x, HIGH);  // turn the LED on (HIGH is the voltage level)
  delay(100);             // wait for a second
  digitalWrite(x, LOW);   // turn the LED off by making the voltage LOW
  delay(100);
}

// the loop function runs over and over again forever
void loop()
{
  if(a*b)
  {
    ww%;
```

You can see that there are multiple lines are written between two backslashes you can write anything but that will not go into compiler will not affect the program at any time but it will create a better understanding of the program for the future you are for another programmer who has seen that you can get something hello world. Thighs are a B LI NK blink programmer and you can now just install it into a bowl you can see that it's installing. Now check if you can see the microcontroller. It's not working because we are stated once. And we have to clear it okay go there and define the function g l or was we have done earlier. define an oldest. Remember to terminate the function after a semicolon. Now install the program And here you can see that it's completed and you can see that the light is blinking.

PUTTING TOGETHER A CIRCUIT

So now you know what an Arduino board and wires are. Time to combine them into a set. This is a centralized picture of a breadboard and I'll explain it to you below. By lugging things onto you to breadboard as as it says just you can have a circuit that will allow you to make an LED light emitting diode blink. So this is is a pretty self-explanatory though it may be a little confusing. So I'll tell you exactly what you what it all means. First let me show you the tool box it requires that you aren't familiar with.

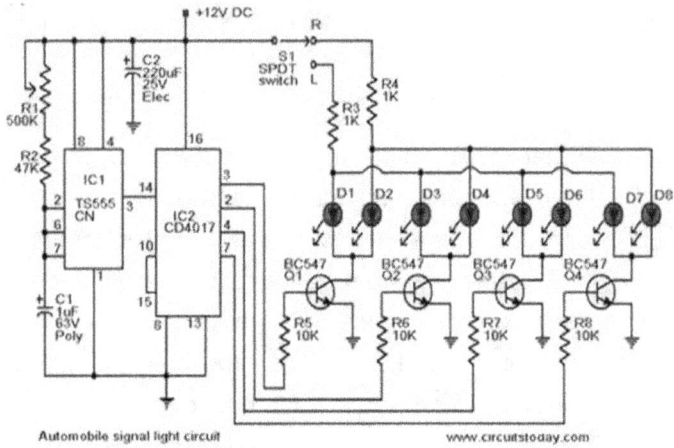

Automobile signal light circuit   www.circuitstoday.com

These are links below not that one of the metal bins is longer than the other. That's important. You should keep it in mind. The other thing is a disaster. Two sisters eyes are kind of complicated. And in order to fully understand them I'll have to explain things like current and voltage. However it's not required to understand circuits use of doing.

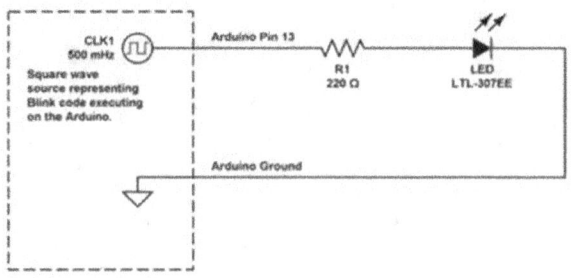

Since that Adreno platform has so much support on the Internet that you should easily be able to find a diagram telling you exactly what resistors you need and where to put it. In order to make a program take it however that one big charge is what resistors look like. Note also that they have lines on them and those lines are also beating on the joint above so that you know which direction the resistors should be installed. So let's bring that photo back. Now you have seen the BART. So basically what's what did steering you to do is to plug your LEDs longer to meet carbon in 2009 and its shorter been into 10 resistor which is three hundred and thirty arms as needed for the circuit then connects brought in to the anode which is minus Cullum. You can't unless the resistor in Annie Hall on the minus column because we really know that they are all connected.

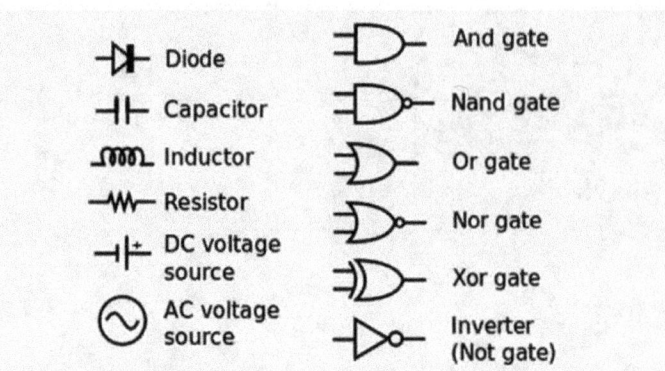

Now just connect the way out does that mean we have a way out from Row 9 connected to Kensington and then I'll on the cathode connected to 5 fault and 1 2. I'm not going to ground not there. Do we know how all the little holes are labeled like on the right is number 13 on the left is the 500. Also in that list there are there is ground. All you have to do is just to block the appropriate wires and so as soon as you plug your wires into the correct haul you need to do a couple more steps to make gullet the blink. At first you have to plug the wire into the Arduino. Then you have to upload a certain card to the Arduino in order to make you blink. Now we will learn how to make that blink. Thanks for watching this lesson. If you have any question please ask that you and I would see you next lesson.

WHAT IS ARDUINO RASPBERRY PI AND PIC MICROCONTROLLER

We will talk about the definition of art we know raspberry by and Big Mike room to a lot.

We'll take a quick look at these three controllers. So let's start with Arduino. What is an art that we know and we know is a set of development boards that come with artistic hardware and software libraries. It means you can buy an Arduino board and start developing your project and instantly the boards are built around the AVR microcontroller is the base software ribose to on the board are written and made available for free coding for Arduino programming for Arduino board is a trial program for an admin on my control. The only difference being that program for Arduino is written in its language called the Arduino programming language. This language is the same from any of C programming language except that everything is ready for us to use the program for reading and writing and. Bit of memory has already been written by someone. We just need to call the functions like it wrong to try and add the address and the value as you can see in this line. Sambo's if you want to blend to that you are the serial bar. You don't need to initialize at registers or boards just call serial. Then try the word hello so that you can print it in the serial port. This is an example of the Arduino board. This is ugly omega as you can see. It has at best a processor. So let's move on. What is my control of big usually pronounced as PIC is a family of microcontrollers made by microchip technology derived from the big 16 50 origins on the floor by instruments microelectronics division area models of Vic had read-only memory or room or field program or even on for program storage? Some with Broad Vision for erasing memory.

All current models use flash memory for programs. And newer manuals allow the back to the program itself. Program memory and that are separated. That one is 8 bits 16 bits and lots. That is 32-bit wide program instructions vary and bit count Pfeiffer many of back and maybe 12 14 16 or 24 bit long. The instruction set also varies by the model with more powerful chips adding instruction for digital signal processing function the hardware capability of big devices that range from six Been us it been the IBM chips up to 144 been as the chips we are talking about a very large number of pins 0 144 been with discrete input-output Benn's ADC and DSC which means convert them to another convert all and communication both such as you are. I do see can and even you be noble and high-speed variations exist for many types. Manufacturing supplies computer software for the development known as beat up. It's not the most famous one and the most famous one called Make Crecy and it's available for free online assemblers and C C++ compilers. Third-party and some open-source tools are also available. This is how the Big Mac. I see Lokes. It's not boarded like here. If we talked about the Arduino SOTT like Arduino board it's just an I see as you can see it's just an icy and you have to hook it up to the bottom by knowing which bins must be connected to or by wire supply. We will talk about this in future lessons. Now let's talk about Rasberry by as you can see this is the spirit by the by is that card-sized computer originally designed for education inspired by the 19th 81 BBC Micro creator IBN uptown.

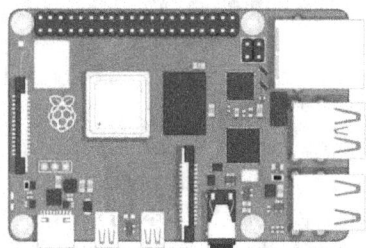 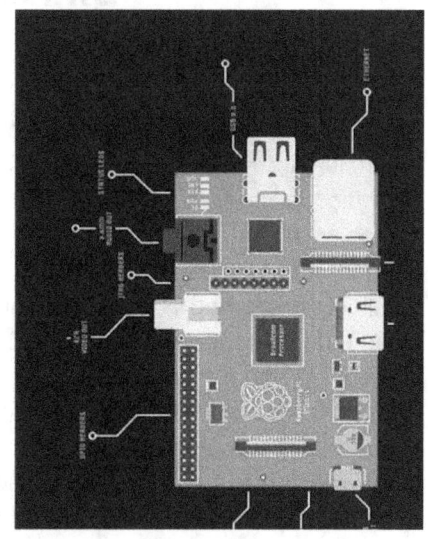

The goal was to create a low-cost device that would improve programming skills and help with the understanding of the university level. But thanks to its small size and accessible price it was quickly adopted by tinkles make cars and electronic enthusiasts for projects that require more than basic math such as Arduino or bedmate controller that I believe by is slower than a modern laptop or desktop but is still completely Linux computer and can provide all the expected abilities that imply at a low bar or consumption level. So this was a quick introduction. It was a fast introduction to Big Mike. Are we not raspberry pi. In case you didn't know any of these. The controllers or minicomputers. Next, we will talk in more details about the difference between these three. But we had to talk about what is each of them so that if you are new come out the last bit about or to my control you will know what is what looks likes what are its specifications. So that's it for this lesson. Thanks for watching. This is education educational engineering team.

QUICK SUMMARY OF THE MAIN DIFFERENCES

The man or the obvious difference between these three controllers. Like my controller Arduino and Rasberry by before going into details let's start with his Big Mac the Big Mac controller from its name. It is a microcontroller based device which means it exile's at controlling small devices. This is it. It's just used for projects like heat-detecting devices motion detectors doors or even a small RC robot. You'll also hear people talking about vintage circuit boards when they talk about Big Mac or not. So when a project is tested on a blackboard and the engineers make sure it works they move to a more permanent prototype called a printed circuit board. As you can see in this image this is a printed circuit board. This is big my controls are on. These are electronic items. It's connected to IBC so you can easily control using B.C.E. or make the controller control the whole process itself. That's forebear as you can see. This is the output. These LEDs will light this motor that will rotate. And this puzzle will send a sound using this big microcontroller regarding Arduino the Arduino, for example, is my controller which means it exile's at controlling small devices like sensors Motors and lights. It's just like a big controller but it comes on board as you can see because control comes on. I see this board has an animal control and you don't buy well you don't buy that controller alone. You will buy the Arduino board in let's say combat or as a full product. This is why the app we know is best used for projects like building. I woke up like a motion detector alarm or even a small but you'll also hear people talk about prototyping with an Arduino which is the process of quickly creating a prototype electronic device. We don't have to hook up any wires or solder items like. And my controller if the prototype is successful and the device works it can be made on a larger scale with printed circuit boards just like with my controller. So you will find that you know my front runner are similar. They have a lot of similarities in many areas. We'll talk about disabilities and you'll talk about the time that you should choose. Arduino all of our Big Macs from the time that you should choose. My control over Arduino. But for now, let's start with this information and move on to Rasberry by the by. On the other hand, is in my control and isn't meant for controlling sensors and other things like that. It's an entire computer with its operating system and is intended to be used as one. The operating system is rather minimal so you'll need some coding knowledge to get the most of it. But that's one of the things that played by is great at helping people learn to code. It's also really good at acting as a server. It can communicate with other computers that serve as an alternative to Chrome cost or provide information and log data. This is how it looks like it has before. If I'm bored has an eyeball so it's I mean the computer has a Sibiu specially created for Rasberry By Thanks for watching this lesson. I hope that you enjoyed the information inside these lessons so far. If you have any question you can drop it in the Q& A board if you want to live. I'm not alone or big. I don't you can't. All fine. We have courses to learn how to make up into a circuit board to learn how to teach you how to make the program Arduino or to program Big to make controllers.

HARDWARE POWER AND CONNECTIVITY-EN

We'll talk about the hardware differences between my Mac on our Arduino Andreas Billyboy when you look at Big Mac from next to us we know next to our spirit by it's very clear that the hardware of the phone is quite a bit between them. Let's break it down. Regarding Bouwer the Arduino Bauer subline requirements are very simple. You can plug it into your computer or a battery back and it will start running code immediately. If the bar is disconnected it will stop. There is no need to run a shutdown process. Beckmann control is very similar. Big Mechelen throttle body supply requirements are even more symbol of volleyballs supply will do their job and bought it only to shut down process. Once done unplugging them supply and that's it the Rasberry by on the other hand because it has a more full-featured computer system or computing system in place should be shut down like a regular computer and can be damaged by Balatka Big Mac. Arduino and that's the spirit by having a very low Bordereau and can be around for a very long time without using much electricity. OK now let's talk about connectivity that raspberry by comes ready to be connected to the Internet. It has a built-in Ethernet board and it's very easy to get. You asked me if I don't get to give it wireless connectivity as well. You can see and this is why. Dongle. So this respited bike be connected to the internet 12 lessly this is one of the reasons why they buy is a device of choice for things like base on a web server blender server and VBN. Been on the other hand doesn't have any built-in capability for connectivity. If you want to connect it to the Internet you will need to add an extra piece of hardware that includes an instrument board. If you want qualified dongle or Arafa connectivity you need a different piece of hardware. Again because the Arduino is meant for hardware projects in the state of software ones it needs a bit of tinkering or tinkering to get it connected. Like no big Mac doesn't have any built-in capability for connectivity. If you want to connect it to the Internet you will need to have an extra piece of hardware that includes an incident board. If you want Wi-Fi connectivity you will need a different piece of hardware. Again if you want to use an SD card or you must be driven with my control you have to build an interface circuit. So that's it for the hardware. Regarding board and connectivity next, we will talk about input and output bends on more interesting topics.

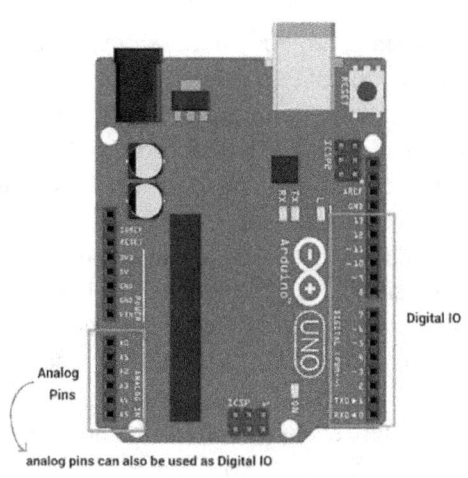

analog pins can also be used as Digital IO

new unison today we will talk about but Outwood bends you us be storage in what is called bends are what allows you or single-board computer to talk to things that are connected to it. For example, we are spurred by good life I lead. Or you could activate a motor or go to pick my controller could fire alarm if you are looking for hardware connections. These spins are what you need that's been hit by two back 17 of these spins. Why are we in on the first 20. My afar. 18 to 68 or 144. And that's I'm with you all depending on the model that you choose. Another significant difference. And then we start with bends between the three boards is the Timberman resolution at which you can control them because the last bit of buy is a phone computer. It has several things that are vying for Sibiu time. Which means it can have some difficulty in getting the timing down to small fractions of a second and it needs software to properly interface with sensors and other devices. The Arduino unbecome I from two, on the other hand, can change the output of the book on their bends down to a very tiny amount of time milliseconds and microseconds. So this is a significant difference between the last hospital by my controller and Arduino regarding storage you know my control that comes with 32 kilobytes of onboard storage which is just enough to store the code that provides instruction for its common program. You can't use this footage for apps videos photos or anything else that I spent it by, on the other hand, doesn't come with any onboard storage. But it does have an across the board so you can add as much as you'd like adding 32 gigabytes of storage will only cost you around $12 with SanDisk micro SD card and you can easily add up to a hundred twenty-eight or twenty-two hundred fifty-six gigabyte. If you're in the last but not least you be connectivity because Bismarckian floor and arguin I mean to communicate with computers. There doesn't come standard with any of us be that you can use for this type of communication. A single board can be used to connect Arduin or Big Mac to your computer via your computer. You must be bored but that's it. You have to make the interface circuit yourself. You have to build it you have to connect it to your controller. While the rest Billyboy, on the other hand, has for you as the boss that you can use to connect to out of a printer on the external hard drive or a wide variety of other devices. So in this let's talk about storage USP and input-output connectivity.

SOFTWARE COMPARISON

Now that we've laid out the differences between the hardware of the Arduino spirit by the Big Mac controller we can talk about software to understand when you want to use on one board or the other you need to know what each one can do and the little that is dependent on the software that complicates or to complicate the issue. The Arduino doesn't come with any software bursar or Blissett it has very basic capabilities to Brett the code that it receives and Alt-Tab the functions of the hardware that it's connected to. But the board doesn't have any operating system or any sort of interface besides the Arduino integrated development environment ideally which is this one. And the right side. What this means in practice is that you'll need to create the software that runs on the Arduino using the ID you'll create a set of commands that the Arduino will interpret and act as simple instructions could say something like 10 that it lights on for three seconds turn it off then the green light on for three seconds turn it off and repeats. You can do much more complicated things but you'll still need to create the program yourself. Fortunately, there is a huge Arduino community that spans the entire world which means that if there is something you want to do with an Arduino someone has probably done it. You can look at their code modify it and make your Arduino do exactly what you want.

This is a great way to learn the principles of coding and prototyping as well. Which is why the Arduino was a great choice for anyone interested in electronics or any newbie and kids or anyone wants to create things or loves to create things. Same thing with my controller. It has a crazy idea that allows you to write a set of course that tells the controller what it should do. In contrast that I spent by come stacked with a fully functional operating system called the recipient. This for us is based on Debian Linux and was created specifically for that body. There are several other operating systems that you can use with the ball most of which are Linux based. But Android can also be installed now in the so-called trust built by Rosby on the desktop. But I think the system for us Billy-Boy Well bulleting systems aren't the only piece of software that the body Lund's. There are also several useful apps that you can use to accomplish different tasks. One of the most common uses of a spider bite is as a media server for which both Codie and Blakes are popular apps code. And next, you can download games server applications calculators and even that I bought Office Office you know the documents, of course, you can't write your book down for us by as well. That's one of the best reasons to get one to them decode by phone. To lead the code and programming language by phone is the recommended language for the by.

But C and C++ Java and Ruby are all pre-installed on the board. While Arduino can be tweaked to Subodh other languages the native language is the best choice.

EXPANDING ONWARD CAPABLITIES

So let's name it expanding on our Big Mac Frontalis Arduino and that's built by our very capable little machines that can help you learn and do other things. But at some point, you are probably going to want to move beyond the basics and try something a little more advanced. This is one of the places where we know and big microtonal shines. Hundreds of chips let you expand the capabilities of the stock board with things like Ethernet and life-I connectivity better motor control speaker and microphone capabilities or touch screen cameras radio transmitters graphical processing and almost anything else you can think of for $20 to $40 of you will your Arduino or Beckmann into something else entirely These chips are called chills and very easily installed. All you do is bless them on top of your and we know and in some cases, big micro-controller sold them and place money can just sit on top making installation IPs as you can see. And this example this is a shield over Arduino board that is supported by is a more self-contained board and doesn't have the same expansion capabilities as the big my controller. There are several hats available that add additional hardware to the buy. However, that gives you some very interesting possibilities. For example, you can add capacitive since all these G-B as touch screens R.G. be panels and even 3D you get self-censored. Then you ask the. Also if you add functionality with dongles, for example, get the wife on connectivity all you need to do is plug in the wife I don't get it. Still, even with these options the spirit body just doesn't have as many options for adding functionality. Not to say that the buy isn't capable. You can still do almost anything you want with it. You just might need to get a little more creative or a shift to Arduino which is also possible.

BASIC PARTS REQUIREMENTS

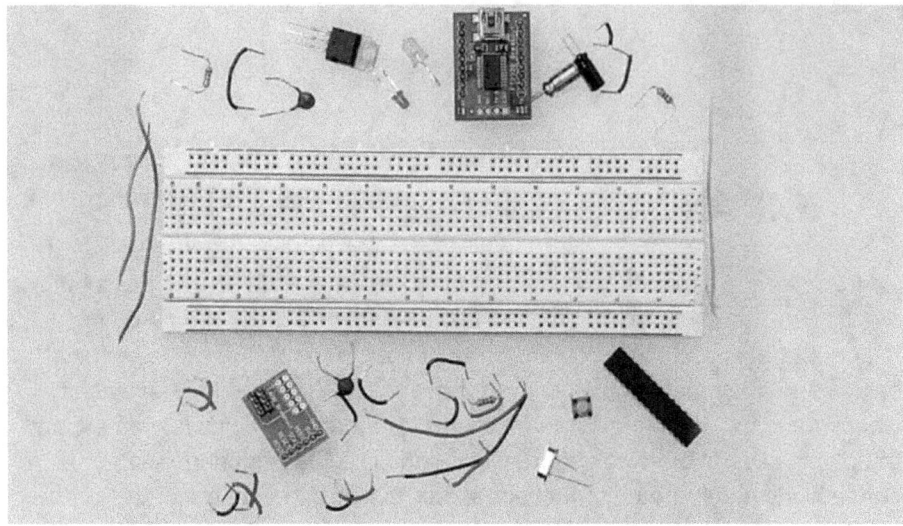

Eight hundred sixty eight three hundred twenty eight. If you are making controller and FDDI breakout board from sport fun or any other store you could also use the new US ASBI Mini will clarify all of these components next. Now let's see the bars that are required for you to make your own Arduino board at home. To do this you need basic bars for wedding bands we know such as a billboard which is this one entity A.W. G. As you can see. These are the whales. Or you can simply use anywhere that is available for you. It's 0 5 voltage regulator which has a 5 volt regulator. This one this is the item 5 will trigger it or takes 12 volt or mind volt provides 5 volt built 5 volts voltage for boarding up Arduino 2 LEDs. These are the toilets for boiler and the ablation to 220 ohm resistors for the LEDs. One thing on resistor for the reset button to 10 micro odd capacitors for that Kristen. These are the capacitors. As you can see 60 megahertz clock Kristen and 2:22 Beco Fraud capacitors. We need also a small that are normally ORBIN off button. For example the Omron type B-3 path all of these components are available on the on line and or flight stores. They are very cheap components. You will make a very reduced cost.

PARTS NEEDED TO GET THE JOB DONE

Old men are bored at home. Now we will talk about more advanced. That must be available for you to beg that board. First, you need to be to serial communication board. You will need a 50 to three to be breakout board. There are two options available from them from Spark Fun or any other online store. There is the after two or three two hour all you US me to see a real breakout board. There is the art we know. Seriously you must be bored. If you plan to use that option and have not yet Soledad Haydar to the breakout board now would be a good time. We will clarify these options later on this course. The second thing that you need is boot loading yours at maigre chips. There are several options for clothing or at mega chips. A few of which are covered in this tutorial. If you wish to bootload your chips using your breadboard and additional balls will make your life much easier but it's not necessary. Ever. Our programming adapter from Spark Fun or any other online store can do the job. We will talk about boot loading and why we need to bootload or make a Ceb in more details later on on this course. But for now, you need to understand that. I see which is the bane of the ball must have code that will help but recognize that it's being used as we know brain so we will burn that code to this. I see that as several options that we can use to burn that code.

WIRE UP A POWER SUPPLY

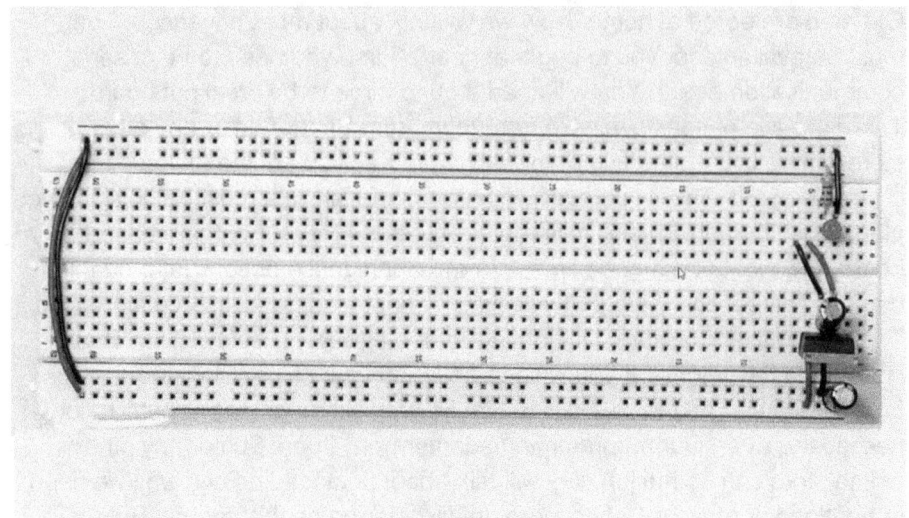

If you have already worked with microcontrollers it's likely that you already have a preferred way to wire up about supply to your board. So go ahead and do it that way in case you need some reminders. Here are some pictures of one way to go about this Fanjul. It was a five vault regulated ball supply. Let's start by doing this at Bauer and ground words for where your voltage regulator will be. As you can see this is where our voltage regulator will be. These are the two main board lines. This one is the positive supply and this one is the ground or negative supply. Here we will place our regulator. It has three bends. It will take nine for volts and produce 5 volts for our breadboard. So here we have our two Tobler lines. Going forward as you can see both our underground wires at the bottom of your board connecting each rail. But on power lines, as you can see we need to connect these two lines with these two bar lines so we need to add two lines to make a bridge the 7 and 054 regulator and the lines to the board. Here we added the regulator and we added two capacitors. The regulator is a.z or 220 bakal. Where is the input from the extent of our supply goes in both in the left as you can see here some extent about? The ground is in the middle and the 5 volts output is on the right one facing the front of the regulator and power out and ground wires that connect to the right and left the rails of the breadboard also are acting like WREFORD capacitors between the end of the regulator and the ground as well as I didn't make that far out on the right trail between Bouwer underground as you can see here. And here we are in the bar on the ground between both undergrounds the silver strip on the copper star signifies the ground leg. So you need to make sure that this silver leg is the ground leg and is connected to the black wire here. You must make sure the same thing cabin's silver with the black wire. These capacitors are used to make a more let's say stable noise-free voltage and pallid and a 220-ohm resistor here on the left side of your board across from the voltage regulator attached to a bar like this is a great troubleshooting trick. You will always know when your board is being bought as well as Quickly know if your board is being shorted. And Blackwells on the left as you can see on the left of the voltage regulator here is where your bar supply will be plugged. So say that you have 12 or nine full board supply. You must add it to these two bins. This is positive. This is negative. It will go through all the capacitors to the regulator. The regulator will produce 5 volts and these 5 volts to the bar line here. Now that it way out is for the board and the black wire is for the ground. Be sure to only attach a bar supply that is between seven and 16. All and know are and you won't get 5 volts of your regulator any higher and your regulator may be damaged. So keep it between seven and 16 volts a 9-volt battery or mindful DC supply or a 12 volt DC supply is suitable for this case. So as you can see here and here. Seven to 16 volt and here are the ground so now that the board basics are done you are ready to load on the chip. As you can see this is your breadboard.

ARDUINO READY

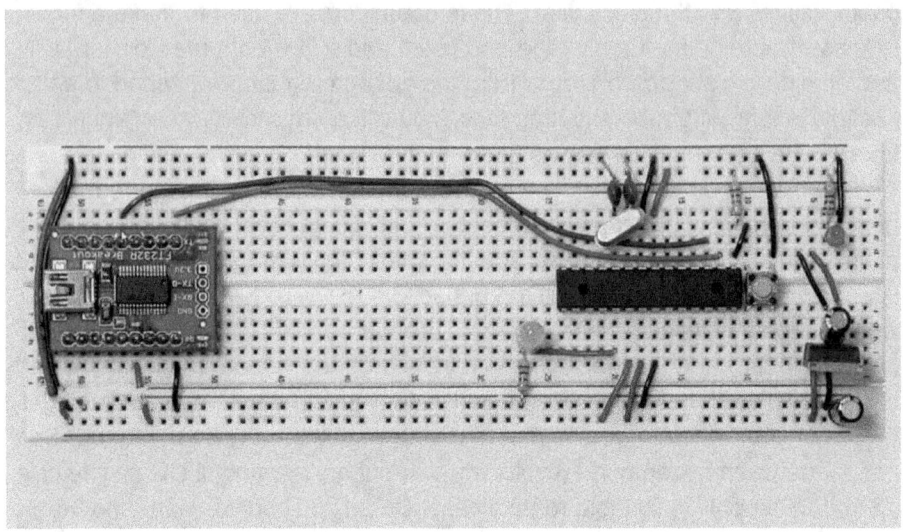

Now we'll be adding ASBI to serial breakout ball to our Arduino breadboard circuit. If you haven't added many headers to your breakout board you will need to do this now. Connect the VCC of the brick by brick outbreed bar the bar on the ground to ground. So as you can see here we have to connect these two bends the bar and ground. After you beat the Syrian border now this is the board. These are the bends as you can see we just connected the VCC underground in the previous lesson on the previous lecture. As you can see here these are the reasons the underground that we connected these are the bends. Curious what all the bends are for the spark fan. After two-three to break up or just simply flip it over. In this situation, we will be using Vcc to supply 5 volts as you can see here from there. Be more to your board ground and the x and x as you can see here is that the X and here is the X we will use these two to allow us to be serial communication. Have been out of the Spark Fun of the break out is this. So you can simply know what each bin does. We'll just use these two pieces the underground to about our we know t s d and r s t to make serial communication via u s p so that you won't have to remove the ball. I see or the chip each time you want to program it. No, it's time to get the speed the serial break outboard. Talking with your new Arduino setup connect door X which has been two of your at mega-chain here been number two to the X of the SB the serial board as you can see. This is been number two. It goes all the way to the tea ex-Penn here and our board and connect to the X bin number 3 which is the black one of your make a chip to the are x of us B move that are unconnected lot x here of the US B to serial board. So we connected the T X and X and in the previous lesson, we can let it slide we connected. He says he on the ground. So now we have our set up a reading and then he will have it ready to be plugged in. Bob and programmed. But wait there is another step-down. If you build you'll make a chip out of your underwear. No, it has most likely been brought down several times by yourself and so it has been bought loaded so you won't need to move any further in this story and you'll have a fully functional breadboard that has a fully functional USP programmable Arduino board. However, if you purchase some extra maigre 328 at maigre 168 chip from an online store there will have not been boot loaded with the Arduino bootloader. Except for the fruit and dust tree if you bought these at makers from the food industry they will have the bootloader already on them. But if you bought these ices elsewhere you will have to make what we call Boot loading. What does this mean you won't be able to program you are abusing as using a serial break outboard and the hardware and software. So to make your new chips useful for us we know you must boot them and must do the boot loading steps.

BOOTLOADING YOUR CHIPS OPTIONAL

It's optional because we clarify why you might need boot loading in the previous lecture booth loading options. There are two options for loading good chips. The first being quite easy and the other being a little more tricky We will cover both so that will be on the safe side bootloader and good at mega chip using our board and on our program. This is the first option. The second option is boot loading. You make a chip in your newly prepared breadboard with and if they are program. So let's talk about the first option. Which is butadiene good a using argued board and on our program, before going further you must know that there are also many different kinds of oddball diamonds but two are most commonly used as you can

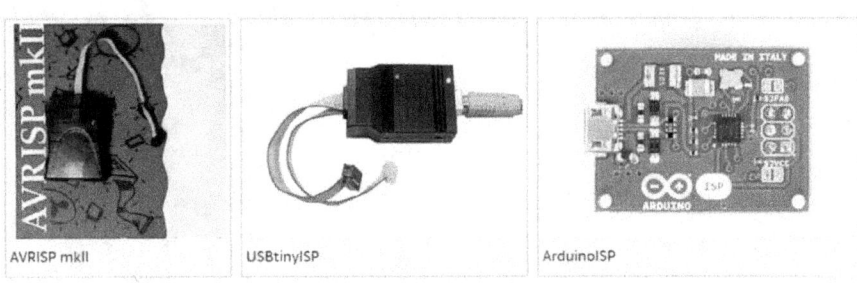

AVRISP mkII   USBtinyISP   ArduinoISP

see in the left we have ever are Ayas be key to this is one. We have you as being beneath us be. I ask B and we have our ISP. These are the three common options. If you are brought Rameau's first bless your mega chip into the Arduino board with the default of the chip facing outward so that you can see here we are blessing or Ceb said that to an extent our board supply and connect a 12 volt Bob rick as you can see we're connecting the dots tabling externally. We are not using They will be more sublime. Please make sure not to use the US B-ball subline your board needs to be X10 Alec Baldwin using the F R I asked B to but is not needed. With that, if you are Tinney ISB then at the six men female Blunk of your arm up to the sixth man are I as she bends with the plastic tub of the ribbon cable heading face inward. As you can see this is our mail hit us six ben and this is the web that comes from that program that we mentioned earlier this every r s b m k too. These are the wires that we must connect to our board here. As you can see and we have to both externally and place our icy here. No the F R S B M K to tens. It's led the way in. When they are being hooked up correctly and are ready for programming then it turns red. If it is hooked up wrong so you must go back. Here is Alade. If it's green then you have plugged up a program of the right way. If it's red then you hope up your program or the wrong way and you have to adjust your hookup or you're set up for programming. This is a very important point to mention when boot loading and make a chip on board if you our programming adapter I scare you from spam.

Fun is incredibly handy as you can see it's a very teeny every hour breakout board. This adapter breakout the sixpence from the arm up to six inline bins for easy attachment to the breadboard. All the bins are also labelled making it very easy to connect it up to your chip. As you can see ground 5 volts on my ass or a c k that said or us-I. This is one of our programming adapters and it's very easy to use the adapter. It's very cheap. You can birches it anywhere. A box or two. Now let's go on and talk about using our breadboard don't want it if you don't have an ever Arbel running adapter you can still boatload without it. It will, however, be more. I've had the urge to set up the two images of the left to the left or the right. In this case, are great references when hooking up to an ATM make a Ceb without an adapter board. The images tell you what all the holes and the sixpence for our blog are. And you will simply need to stick wires in the end and drown them to make a chip. As you can see the top few. This is about all of you.

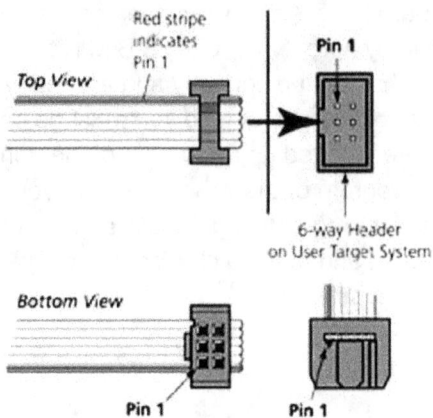

This is the been six-way hither and the use of that cable head bends as you can see on my ass. Oh, I see. Care to set VCC ground. Or us II. These are the sixpence. This image is viewed from the bottom and labels issue of the holes. Take note of the square as to what orientation your cable is and as you can see this will show you the orientation. This box here so that you won't label the spins the wrong way.

FINISH UP AND BURN THE BOOTLOADER

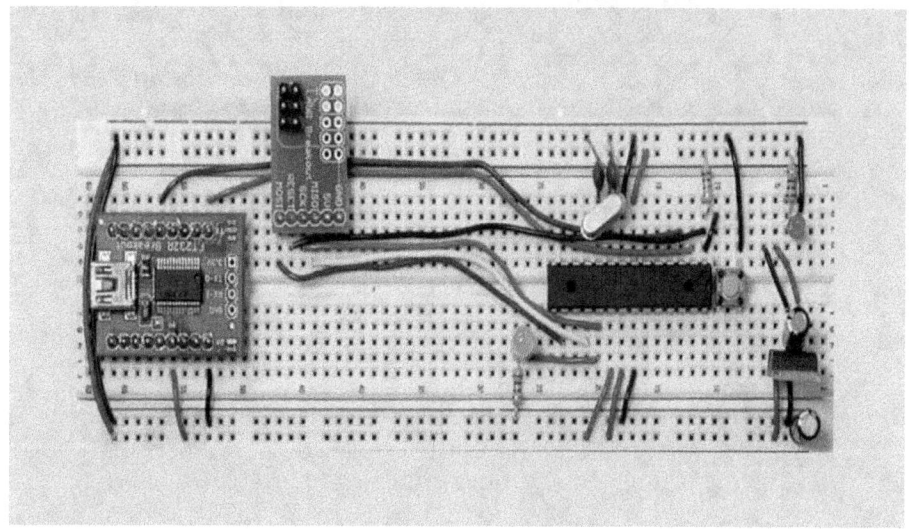

Now Blug there is an audible drumming adapter as you can see here into the breadboard with the ground been matching up with the ground why are you just thrown on the fivefold. Been catching up with the bore. Wow that was just from this is the adapter that we are using moving forward. And this will lead to the last four wires needed by the ever. Our programmer for robot mode loading. Be sure to refer to we my being for helping wiring this up dumb ass Auban of your adapter will go to be an 18 or 20 digital been 12 of your make a chip depending on which make a chip that you are using the cabin of your adapter will go to 19 or and we know there's still been 30 of your mega chip that said Ben of your adapter will go to one of your nigger chip them or us. I've been a few adapters will go to pin 17 or there's still been 11 of your I can make a chip now. These are the wires that they are connected to Andrina chip or make a chip almost there just plug in the cable or you might be a breakout board here and plug the six ben Blog of your if are this one. Or if you are an adapter the black knob of the six bin head must be facing upward towards that mega chip in the next table will show you how to use the software toolbox to burn your bootloader. So this is a simple Sibyls out without using an actual arguido board just using every programmer and a breakout board.

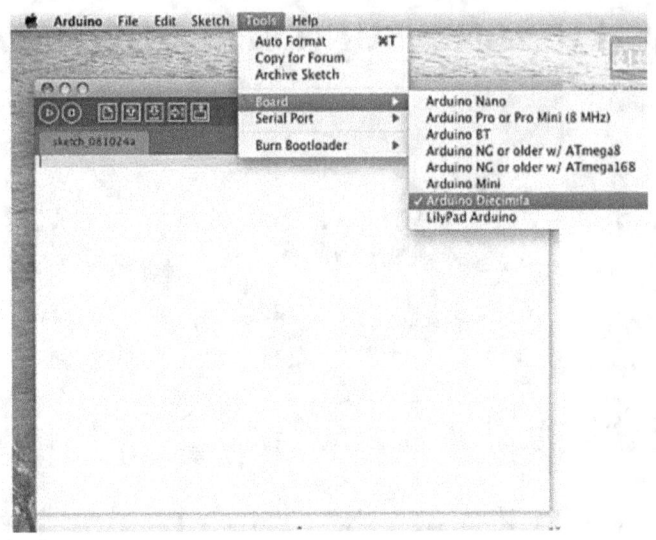

Time to burn Fire Arduino and then go to tools as you can see then choose your board choose the type of Vold you would like to use will effect which will no doubt you will be both on your chip. Most commonly you'll be using that this says the smell area or the most recent version of Arduino for that may be the IAB. However, if you'd like to boot load an Arduino Linley bad Arduino many Arduino or any of the older Arduino versions. Choose the appropriate board from here. We recommend this option. No arguing with the senior after choosing it to go to Tools Brande bootloader and choose the program or you'll be using in this class we are using every hour I'll be ok too. We just need to choose it. Once you choose your program out of our program or we'll begin loading good I can make a chip and a message will appear in the status bar here. Which leads burning bootloader to input out the board. This may take a minute light will flicker on your program and then you will see this message done Bender man bootloader. When done with clothing the status bar will be updated with the message done and bootloader. Your chip is now ready to be programmable using to be programmed using the software. Congrats bar cycle your Arduino and your new mega chip will be running a similar blink program with Ben 13. If this is not the case try and get with one. If this is working was most definitely a success and you just made your own Arduino fully functional board, not on occasion. And I make a chip with the ISP M-K to Celtic and extorting Nungarrayi a long period used when it should only take a couple of minutes. And there are Tinney ISP fish is much quicker. However, there are times where after 5 to 10 minutes it still appears to be both loadings. I found this to be or the Habs. It's terrible checking the data flow and after giving it both time 10 minutes or so I usually unplug verbal or only to find the building block has to be a success and has ended long ago. I by no means enduros this. So and you take all responsibility in whatever you may have been to your chip. But in my experience, it has been a success.

ATMEL ATMEGA 328 DETAILS

The ATMEGA three to eight details well we have talked about earlier the advantage to it is the black dollar chip on the board. Thighs are known as Atmel ATMEGA three to eight p p, p q because it runs on five volts and it is power saving chip.

ATmega328 Features

Features
- High Performance, Low Power AVR® 8-Bit Microcontroller
- Advanced RISC Architecture
  - 131 Powerful Instructions – Most Single Clock Cycle Execution
  - 32 x 8 General Purpose Working Registers
  - Fully Static Operation
  - Up to 20 MIPS Throughput at 20 MHz
  - On-chip 2-cycle Multiplier
- High Endurance Non-volatile Memory Segments
  - 4/8/16/32K Bytes of In-System Self-Programmable Flash program memory
  - 256/512/512/1K Bytes EEPROM
  - 512/1K/1K/2K Bytes Internal SRAM
  - Write/Erase Cycles: 10,000 Flash/100,000 EEPROM
  - Data retention: 20 years at 85°C/100 years at 25°C[1]
  - Optional Boot Code Section with Independent Lock Bits
    In-System Programming by On-chip Boot Program
    True Read-While-Write Operation
  - Programming Lock for Software Security
- Peripheral Features
  - Two 8-bit Timer/Counters with Separate Prescaler and Compare Mode
  - One 16-bit Timer/Counter with Separate Prescaler, Compare Mode, and Capture Mode
  - Real Time Counter with Separate Oscillator
  - Six PWM Channels
  - 8-channel 10-bit ADC in TQFP and QFN/MLF package
    Temperature Measurement
  - 6-channel 10-bit ADC in PDIP Package
    Temperature Measurement
  - Programmable Serial USART
  - Master/Slave SPI Serial Interface
  - Byte-oriented 2-wire Serial Interface (Philips I²C compatible)
  - Programmable Watchdog Timer with Separate On-chip Oscillator
  - On-chip Analog Comparator
  - Interrupt and Wake-up on Pin Change
- Special Microcontroller Features
  - Power-on Reset and Programmable Brown-out Detection
  - Internal Calibrated Oscillator
  - External and Internal Interrupt Sources
  - Six Sleep Modes: Idle, ADC Noise Reduction, Power-save, Power-down, Standby, and Extended Standby
- I/O and Packages
  - 23 Programmable I/O Lines
  - 28-pin PDIP, 32-lead TQFP, 28-pad QFN/MLF and 32-pad QFN/MLF
- Operating Voltage:
  - 1.8 - 5.5V
- Temperature Range:
  - -40°C to 85°C
- Speed Grade:
  - 0 - 4 MHz@1.8 - 5.5V, 0 - 10 MHz@2.7 - 5.5.V, 0 - 20 MHz @ 4.5 - 5.5V
- Power Consumption at 1 MHz, 1.8V, 25°C
  - Active Mode: 0.2 mA
  - Power-down Mode: 0.1 µA
  - Power-save Mode: 0.75 µA (Including 32 kHz RTC)

Okay, the frequency well it has a built-in clock that runs on eight megahertz, but by adding a crystal on it, you can see that here thighs is the crystal and after adding a crystal on it, we can make it run or something stain megahertz and all the Arduino boards with ATMEGA three to eight work on 16 megahertz processing speed. And the next part is flash memory or you can say the simply the hardness of the microcontroller. It is a 32 kilobyte, and I'm investing 32 kilobytes it means that it cannot store photo or other data but it is made just for storing the program that is written on the chip. The program is written on the chip in the form of hexadecimal, but we have to program it in the Arduino IDE with c++ language. Our will Arduino is a special form of c++ language that is very easy to use and it can be used for non-professional sounds about the pins. It has 28 physical pins, well you can count here it has 28 more than 28 pins because the ground code has been given two times and also three one through 313 power adapter is also added on the board and three pins for input-output reference. Okay? The next points digital prints well on board you can see that it has a straight 30 digital pins. Well, I want to tell you that first of all consider that print zero and pin one as nor should we use for the processing because if we are doing a serial connection and then if we are using pin zero and pin word there, then there will be an error in the serial communication so we have to prefer that we don't prefer pain zero and pin one for our programming and other purposes. It has five analogues pins such as a zero a one a two a fives well all the analogues pins can also be used as GPIO digital pins bus, but not that point that pins a zero should be written as detail with 14 and a one as a digital pin to pin and so on.

ARDUINO SOFTWARE FOR THE FIRST TIMERS

The next module Arduino software for the first time. In the thighs module, we are going to discuss Arduino and what is a red board. Let's get started with programming download and install of the software in about the Integrated Development Environment and writing our first program that is blinking an LED or you can say hello world. Before going there. I want you to go to the chart to go to Google and search for Arduino.

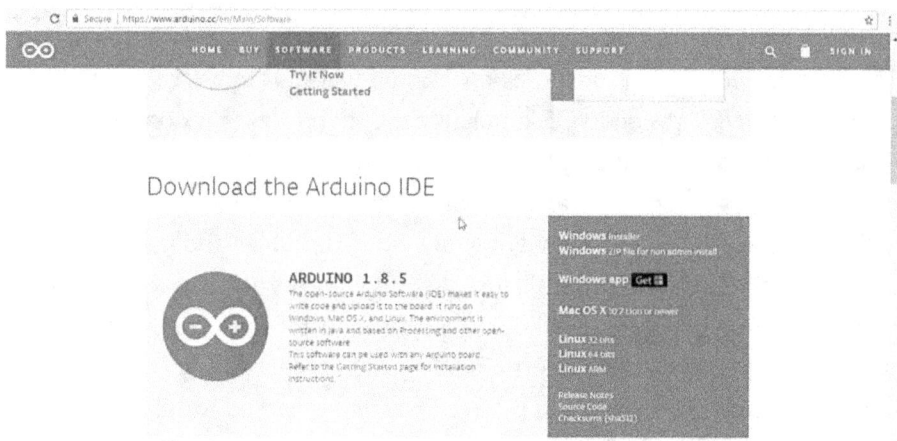

After searching Arduino, you can click on Arduino button Arduino home and after doing that, just click on the software after clicking on the software you have to select your type of installation. It could be a window It could be a Mac or it could be a Linux for me I have selected Windows Installer after clicking on that you will get to the next page that is telling about contribute to download or just for our case right now we are going to just download after the When the download of the Arduino software will be in well I have downloaded it earlier. So you can just see here how the installation of the Artemis okay. Thighs are Arduino 1.8 point four that is for Windows.

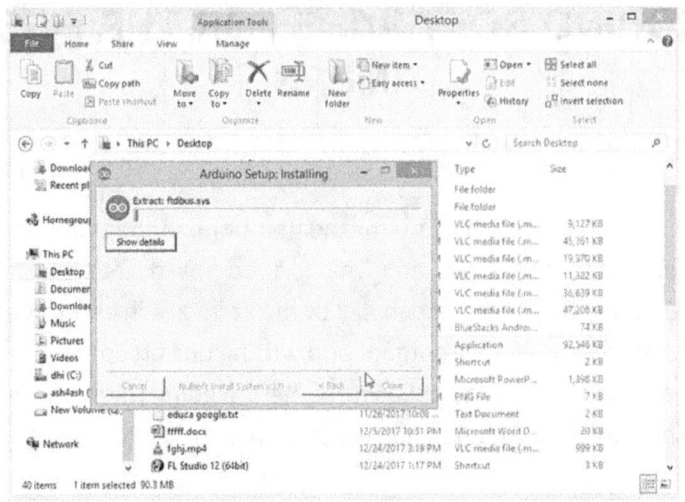

So it is an executable file. Just click on it and it will start to ask the security question just click yes on it. Click on I agree. Click on Next click on the location it will ask for the location where the agenda should be installed. For me, it sees program policies and the folder name Arduino. After clicking it, it will just begin to install. If you want to see how the things and how the libraries or files are going to stall, just click on Good as you can see here, well within a minute it will get installed. Well, I have just it will ask for the driver that should be installed in the Arduino. So just closing it. Okay. Now on the desktop, you can see that there is an icon, which is named Arduino. Okay, just see there. If you don't find the icon, don't worry. Go to C. Go to Program Files x86, click on Arduino and there you can find the icon that is Arduino e xe just create a shortcut or have it on the textile and you can begin Okay, click on it will ask for the access for the Java file click on Yes. And now you can see the rd. Thighs are the integrated development environment for Arduino here you can see that it has all the features that are needed for us for the board if you can garden of water you can find here the port and all the necessary buttons.

DIGITAL READ AND SERIAL PORT

Okay, now, about the digital signal. For that, we have to set the pin of the Arduino as input so that it can read it. So can it read the voltage Again, we have to open the serial monitors we have to initialize it so that we can see the state of the light? Okay, let's just go to Arduino Okay, let's just go to Arduino open the Arduino panel Okay, go-to example select basics and select the retail rate cereal, Okay, in thighs program it is written that first of all the value that the pin that we have to read it is said as PIN two so go to PIN two and connect a wire on it Okay, in the SR world setup you can see that in the word setup, you can see that first of all they have initialized the serial monitor by setting the board to 9600 okay and they have the pin mode or the push was done you can avoid writing push-button data, just write the PIN such as PIN two and settings at the input so that it will read. Okay, now go to void low.

It has said That old button state will be digital read push button the modal push Verdun and write it value to for your better understanding and a clear knowledge of the thing in the world loop the program in the void loop the program is saying that the button state variable will check the state of the data into whenever the state of the data pin two will be changed from zero to one or vice versa. It will give value to button state and the button state will be printing it and the next line you can see that it can it will be checking the microcontroller in every my one millisecond. Okay, first of all, upload the code to the microcontroller. We are uploading it will take some seconds to compile it and then convert into hexadecimal and download it to the board and now it is converted open the serial monitor. Ok, now we can see a garbage value or a value that is changing from zero to one Don't mind that it is only garbage that is generated by the microcontroller.

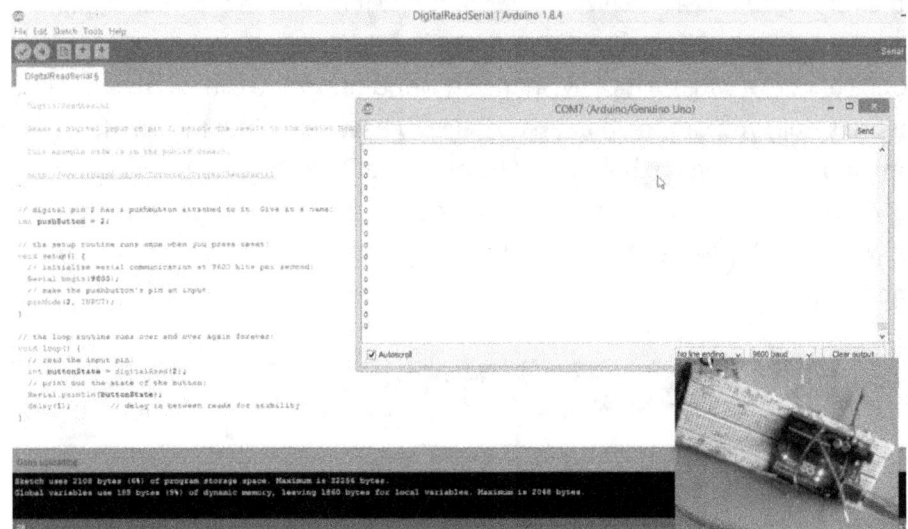

Okay, Here you can see three wires. First of all the wire that is connected to the digital pin two seconds a red wire connected to five volts and a black wire connected to ground. Whenever you connect the digital pin to the black wire it will show a fixed value of zero. Okay, thighs are saying that the data value has a low value or it is connected to the ground. Whenever you connect it to the five volts it will show a value of one again let me show you the day the pin that is connected to the digital pin two okay connected to ground. It will show a fixed value of zero. You can see that the garbage after garbage you can click it you can again see that it has a fixed value of non changing because it is set to ground or it is set to low. Okay, another time latest connected to the five volts and you can see it has a fixed value fun. The moon will be causing garbage we don't want the garbage but whenever we connected it will also fix value Fun. Well, thighs are the program for the data pin and thighs is for the digital read in the serial monitor.

IF CONDITIONAL STATEMENT

We are going to talk about one of the most important things in programming that is the conditional statements about the conditional statement, it can be simple if statement.

Reference > Language > Structure > Control structure > If

if...else

[Control Structure]

Description

The **if** statement checks for a condition and executes the proceeding statement or set of statements if the condition is 'true'.

Syntax

```
if (condition)
{
    //statement(s)
}
```

Parameters

First of all, it will check the condition is a Boolean expression. It means that the condition that is been declared into the two brackets if the condition is satisfied, then only the in it will go to the statement if the conditions are not satisfied. It will not enter the loop or in other cases, it will go to the sloop. Okay. He alleges C and n. First of all, let's study about it, okay a condition statement in programming is used to define a condition where the condition is a Boolean expression, it can only be activated when the condition that is declare and condition that, of course, admits if the condition is different then it will not occur, when it will go out of the loop and maximum time it will stay in the loop. If a statement can be stand-alone It means that it does not need an L statement, while the, if statement with an L statement can possible, can be possible but individual a statement is not possible because it needs an if statement with it to them. Okay. But the most important thing about the condition is that it's a possibility to become a hierarchy.

If statement

The if statement can be used in a branched manner Such as when the condition 1 is fulfilled, it will take the other condition into consideration and fulfil the required condition 2 if the parameters are satisfied

```
    [ 1 ]                      [ 1 ]
        [ 2 ]                      [ 2 ]
            [ 3 ]                      [ else ]
                [ 5 ]
        [ else ]              Independent conditions

Inter dependent conditions
```

You can say that in thighs case first of all the first condition is satisfied then it will go to the second condition. If the second condition is satisfied, then it will run the statement. I never In thighs statement the third condition is returned then it will go to the third condition and thighs thing you can see that it will go to the most insert condition that is placed in the hierarchy. But if the first condition is or the very point is not satisfied the program will jump directly to the statement while the conditions can be placed independent of each other without any higherarchy you can say that if five occurs it will go condition where it six occurs it will go condition to any anything other than five or six occur then it will go to the is condition. Okay, now let's just see a simple example of what is conditions and how they are none. Okay, go to Arduino. Make a new document. First of all, we are going to write about our LED pin that is been connected to PIN nine.

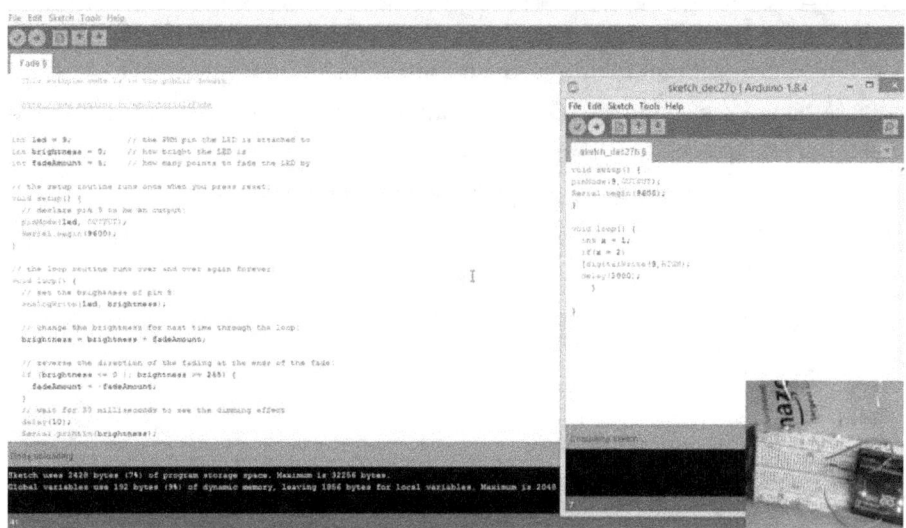

Okay, first of all, it did thighs right pin mode murder and death The pin nine as output or UT t UT pin nine has been declared as output okay but we want to see the conditions occur more or better with a vision of the condition we are declaring serial also we have declared cereal in giving it aborted of 9600 Okay, that's done now go to the loop our void loop okay in the void loop first of all for a condition we need some integer we will declare it ind integer x Okay. Now go to the conditions you can make a simple statement with the condition such as if x is equal to two then thighs is the condition and there it has been declared. Go to two curly brackets and write something such Digital w ra te digital writer PIN nine number nine s high okay in the program, first of all, we are declared PIN nine as an output we have called the serial monitor after that we are taking an integer and giving a condition in the if statement says that if the integer x is equal to two then only it will go to the statement that is saying that right digital nine is equal to high and you can say that it will be a good deal it will run for about two seconds. Okay. But if the loop is not exited, it will go back and it will be running continuously. Okay, let's just take it. We are here we are going to a value of one. Okay. Uploaded it first compiling the sketch and it's going to be Within a few seconds it's uploaded you can see that if x is equal to one Well that was a syntax mistake I have to give it to equal to okay now, let's just see we have done a small mistake and I will be explaining about that okay first of all if you want to say if equal to then you have to always give to a value of two equal to science, if you have given it a one equal to send a duty saying that x is told a value of the given on the right side it means that x is storing the value of two, but if we are giving a to equal to sign it is saying that x is comparing its value to the right side integer that has been given okay else it is not blinking at any car any function. Now you just give the integer value to now run it. compiling and then you can see that the LED will glow okay.

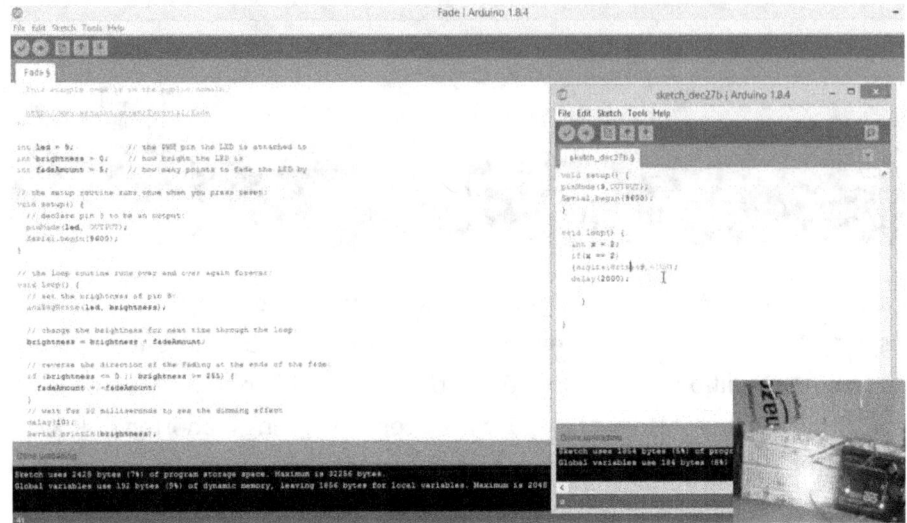

But if Want to see the state How can we get a clear declaration without giving it a state? Go here in the condition and read serial dot write conditions, okay. Is serial dot t ri NT ln ln is important to write it to the next line after the first line has been written, we are giving x. Okay, let's just again compile it and upload it to the world.

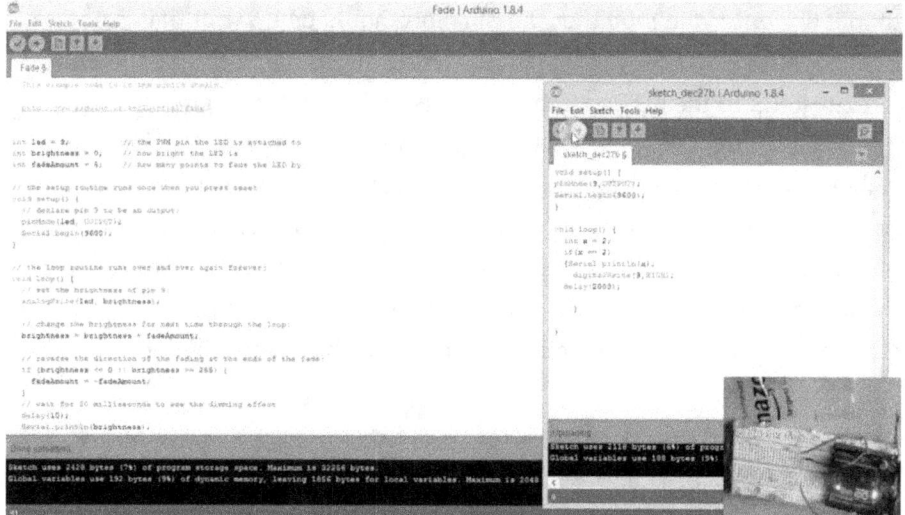

We uploaded it and ledger CNC then monitors the value of x is equal to two. Okay, now, about the L statement. You can just simply write e l SP without any braces or anything goes down write two curly brackets. digital writer. nine is equal to let's just go there and do an exponent.

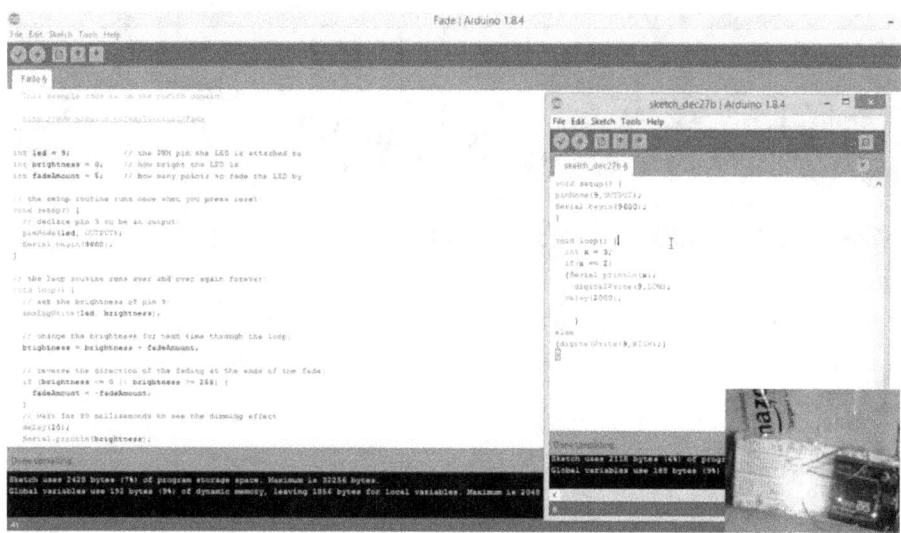

We are getting it High thighs time Thighs will be high and if the value of x will be equal to two then thighs will be L or w low and we are going to change the value of x is equal to three. Now in thighs program, we can see that we have declared the value of x is equal to three, if the x value is equal to two, then it will enter the loop that the pin nine should be low but that is not possible here because we have set the value of x equal to three so it will go directly jump to the else statement who is saying that the light will be high or it will be glowing Okay, let's just compile and upload it we are uploading it will take a second now you can see that the LED is glowing and if we want to see the value of x, you can go to the serial monitor and check here the value of x is equal to okay there is a mistake that I have declared the value inside the if statement so the serial print will not occur because inside the statement and the statement is itself not occurring. Okay, go here.

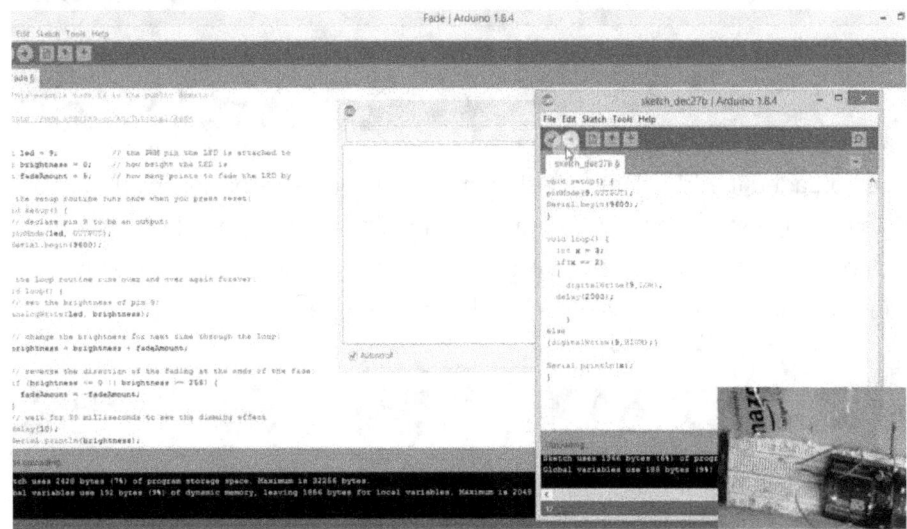

After the statements both the statements after have been terminated write the value of x it will be always printing the evaluator condition has been met or not. Okay, let's just check it's given the value of x is equal to three. Now if you say there the value of x is equal to two, then it will enter the if statement.

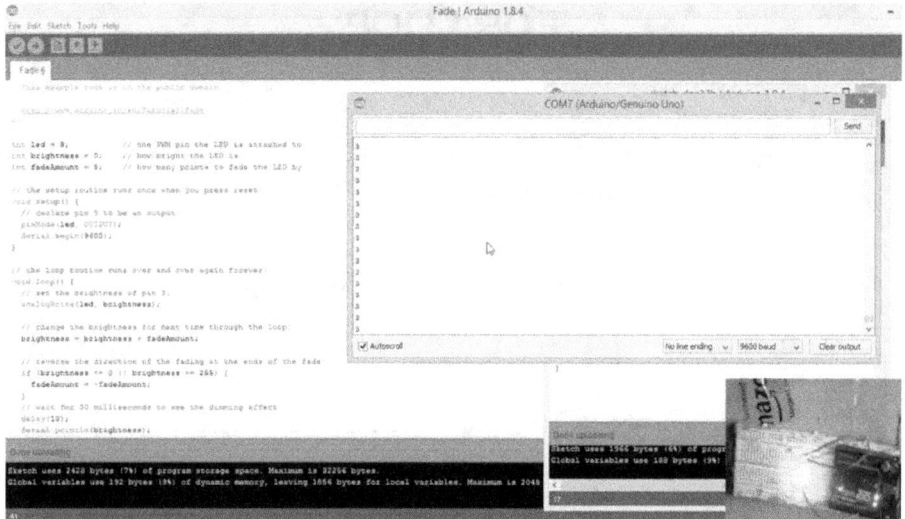

Okay let's just check it you can see that it will be taking it is taking some second let us give it a restart okay you can see that the light is not glowing and the value of x is equal to two. So that is has entered the issue. It has entered the if loop and in the flow, we have declared the statement that the light will not blow.

FOR LOOP

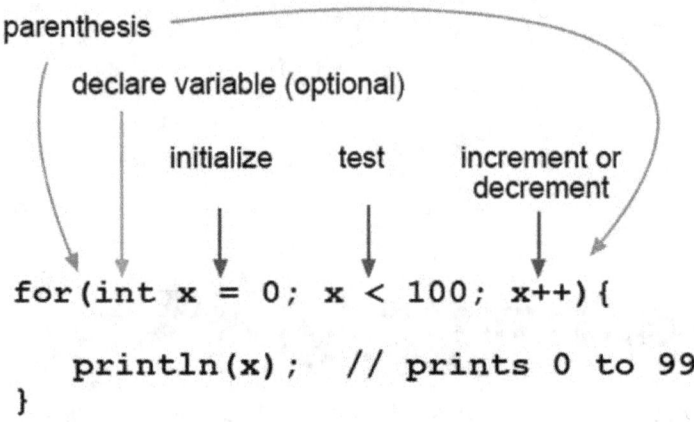

```
for(int x = 0; x < 100; x++){
    println(x);   // prints 0 to 99
}
```

Well for loop is an important compound of programming that has been used in almost all type of programming about for loop is used to run a specific type of code s. For loop is used to run a line of code for a specific period or to perform a certain interval or taking certain objects. After that, it should be exit performed the condition of exit Okay, let's just try to write a for loop in the owner rd not to get a physical computing example of it. First of all, for loop is used to write a statement for a political interval of time then after the should be exited. There are two methods of exiting, first of all, is default without writing anything else. But the second method is Important It is called the break of the loop we can break the loop at any time as we required. We are going to discuss both of them right now. Okay, let's just, first of all, open the Arduino, open a new sketch and ask for thighs method.

```
int x = 0;
void setup() {
  // put your setup code here, to run once:
  pinMode(9,HIGH);
  Serial.begin(9600);
}

void loop() {
  // put your main code here, to run repeatedly:
  for(k = 0;
}
```

We have introduced an LED at PIN nine. So we are going to do our work on that. Okay, let's just write some code. Pin mod. m should be capitalized because it's case sensitive. More led nine. HIGH Thighs is in the brackets properly. Okay, and open the serial port. For a better understanding of the situation. CDL begin 9600 sets the baud rate, Okay, now come here, thighs are the part where the loop plays the role okay right for is thighs is the word for declaring the forum, give two parentheses. And let's just see the syntax, First of all, if we want a variable and if we want is to get a fixed value and it should be open everyone, we should open for every situation, we should always make a variable a global variable.

```
int x = 0;
void setup() {
  // put your setup code here, to run once:
  pinMode(9,HIGH);
  Serial.begin(9600);
}

void loop() {
  // put your main code here, to run repeatedly:

  for(x = 0;x<5;x++)
  {digitalWrite(x,HIGH);
  Serial.println(x);
  delay(1000);
  }
}
```

If we write it in the word loop, it will be a local variable, but if we write it before the setup and before everything, just declare ain't x is equal to zero now it has become a global variable and its value will be taken only for the first time. After that, in every condition the value will be changed as for the loop, okay, in world low, we have written the for loop and we have given it parenthesis but now lets us do the work. Excellent. Call to zero, give a semicolon it says that the statement in thighs condition was given x is equal to zero, x is less than, say five. Now x should be increased. Thighs are the method of writing for a loop. First of all, declare it give two balances record, write the value of the variable v one, x is given the value zero. After a semicolon, we have given x is smaller than five, as we understand them, increase the XOR to one step, as we want we can also write x is equal to x plus one, but it is simpler to write x plus four giving a single increment every time get two curly braces, open them and write the ID at L digital writer. Exhale, comma. HIGH, it means that for the loop when x is smaller than five, it will give you a digital right pin value. have high Edmonds and the light will be glowing de la Vida de la should be introduced for a better experience of the program. And it is important in most of the cases to get a better perception of what we are writing and what was the program because the processing speed of micro gondola is very high, but we want to control it as per our work. Okay about the word we want to get on the serial monitor.

declare it is L serial dot printer I am using always print ln to get the line and get it to work on the next line. Correct thighs index serial dot printer

and we are printing the value of x on the screen. Okay, that's cool. Four x is equal to zero x greater x. Four x is equal to zero x is smaller than five x will be incremented every time before reaching five.

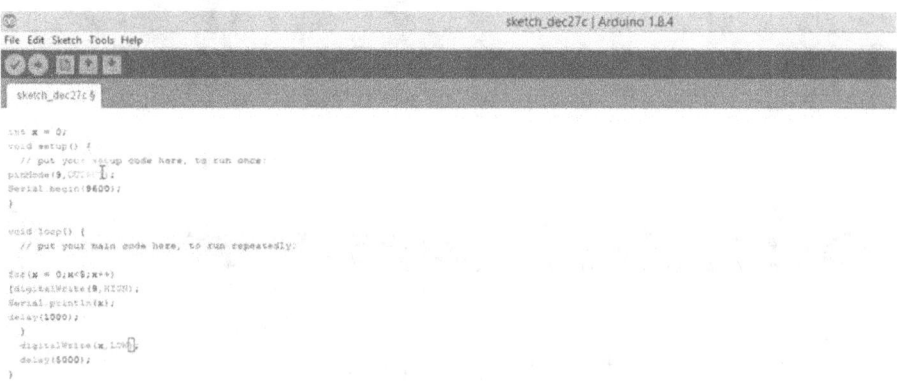

okay here now what should be After the loop after ending the loop we are given the value digital writer at low but the process will be very fast because the speed of the microcontroller is very fast so we have to delay it for next five seconds okay let's just say the light will be off for next five seconds give the value 5000 here and terminate the line okay. We have declared in x is equal to zero void setup we have called led nine is equal to pinaud LED nine is equal to output all OUT, PUT everything should have been capital we have called serial begin 9600 okay in Word log we have made x is equal to zero x less than five x should be incremented by every step and incrementing every step everything the return statement will be running Okay, the data light at pin nine will be high for next five seconds and the value will be printing on the screen. But after getting it high on x is equal to greater than or equal to equal to five, it will terminate the loop it will exit the loop and go to the next statement that is showing that the pin nine should be low for next five-second, okay, let's just compile it.

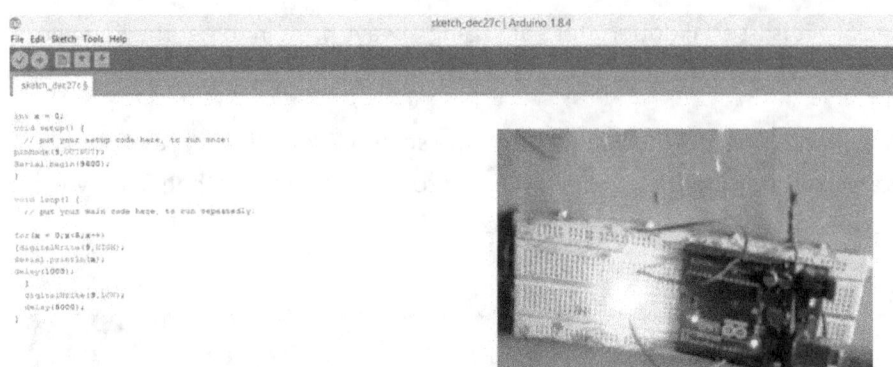

Taking some time, no problem. There are no issues. Okay, let's just upload it to the world. And uploading is done. Okay, let's see the serial monitor is better for better understanding. The serial monitor is open in the backward Okay. Let's just get it now. Light is high for the next five seconds and after entering five it will be low for the next five seconds. Again it will be high and again it will below. Let's just say a brighter export a better experience of the four low we can just introduce a value known as a break. If we want to break before five, let's have three.

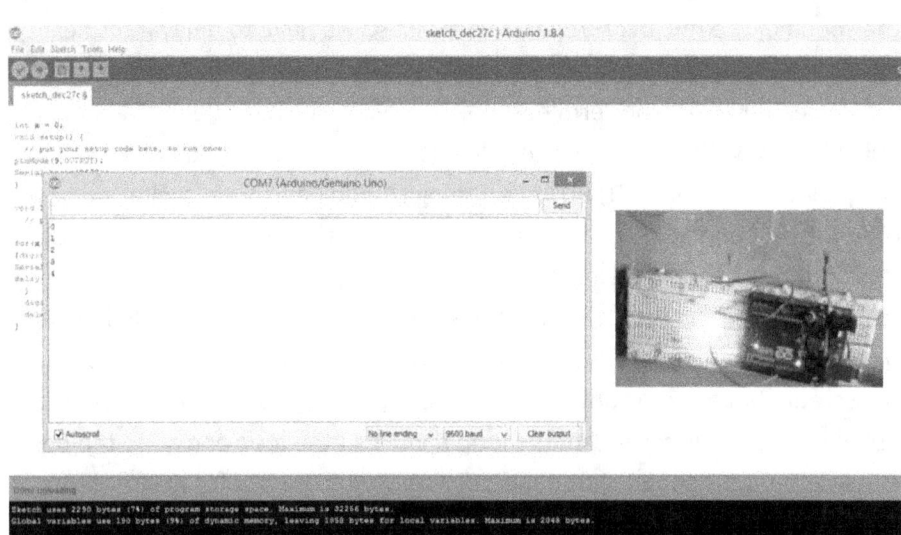

So we can write equal x is equal to three do the curly braces right break b r e AK break is the method for exiting the loop. We have written it in the for loop,

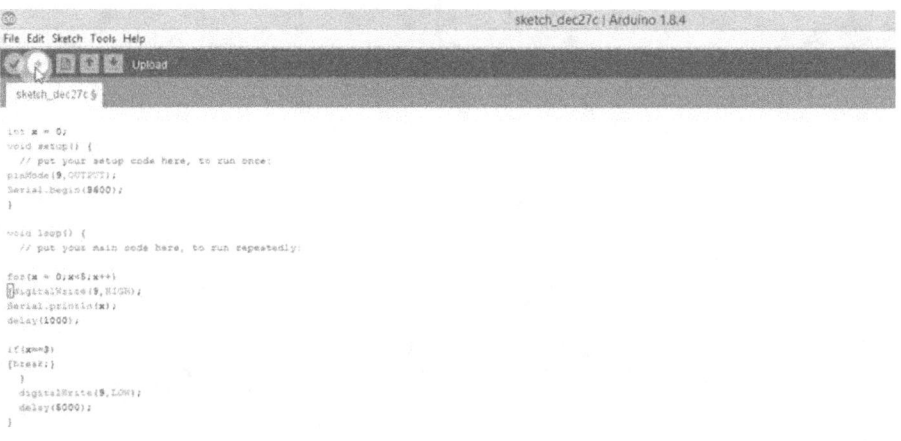

see that in the for loop we have written x if x is equal to three, then it will be directly jumping out of them for a loop. Okay, let us upload it to the world. For loop, I'm saving the code. Okay, it's uploaded on the board. Let's see what happens it's uploaded and the upload is done properly open the serial port, you can see that on three it will exit out of the loop okay you see that after getting on three, it just exited out of the loop because of the break statement that we have included. It is a very important statement for the writing program and getting the performance as far as per hour in a bit getting the performance as per our requirement and thighs is all about for loop is mostly used for calling the array values because it is very important to get a list of a read that has been specified each value at each step but the best method to call an array is for we will be discussing more arrays.

WHILE LOOP

while() loop

```
             Condition
                |
while (digitalRead(2) == HIGH) {
    Serial.println(x);
}
```

Loop conditions are same as "if conditions"

While loop well while loop is similar to for loop as we have studied previously, but there are some basic differences between the for loop and while in while loop it does not require an increment, while in for loop the syntax says that it requires an increment and four using a complex statement even you can use a Boolean statement to run a while loop while thighs are not possible in for loop. Another important thing that a for loop is a very good example for calling an array while a while loop should not be briefer for such experiences. The basic difference between a four loop and while loop is that a for loop is specifically simple and controllable loop while a while loop is individual and it can run on an infinitely long time and it can also use the Boolean values as conditions it can require it. It can also use Boolean values as the condition that are required To get the statement Okay, let's just perform a while loop example here.

If you want to learn the basics of the while loop and how can it be used for different complex programming example I am showing you a website which will give you a basic difference between them. Now you can see that here it is returned that for a while loop the conditions and the for loop the condition that is met. Okay. Let's just go back and open the Arduino and let's just write a code simple code for violence. As usual, I have given an integer value x and it is a global integer because it is a global variable with the entity as declared before, I am giving a variable a global signature so that it can be operated from all other the part of the programming and later Just say that we have connected an LED to PIN nine okay.

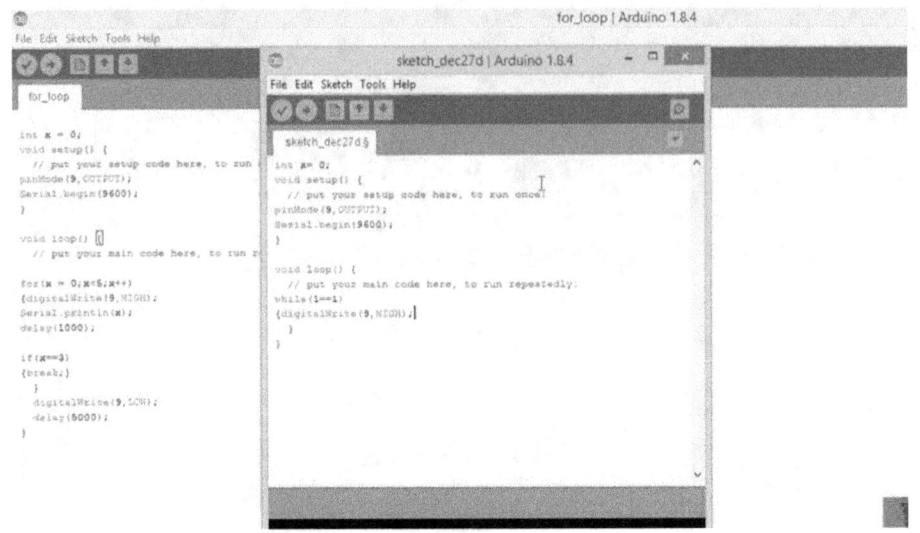

So declare it pin mode B mode nine comma output should be capitalized as I told you earlier and open the serial monitor for the better understanding. If the word does not turn orange Thighs means that you have made a simple mistake or anything such as capital or making alphabet lowercase or uppercase or anything any syntax mistakes is done then the word will not change to their default colour for example, for calling something it will turn into orange said awarded that is fixed for all tests. We will be performing another board example in the lectures. Okay, now get into the while loop is initialized calling simple while you can say One is equal to one thigh is the most common example of calling the while loop. It will make the LED glow for infinitely long time. Okay you can see that here if one is called one that that is a basic knowledge of mathematic that is always true, then it will run thighs is an example of the Boolean statement that the while loop can take inside it. Okay, let's just upload it to the world it's uploading Okay, you can see that the light will always glow because one is always equal to one and the T thighs is a known fact.

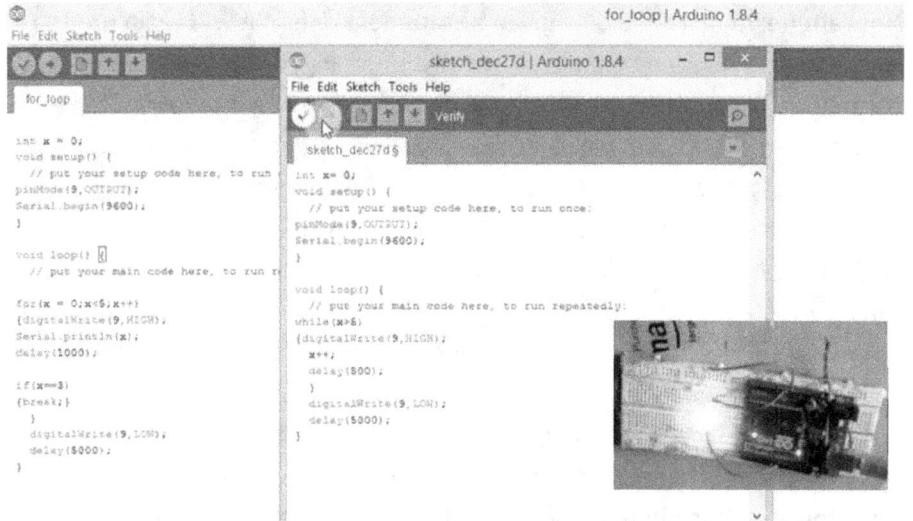

Now, in while loop you can perform other statements also such as while you can say that while x is equal to while x is always greater than five, thighs are the best example. You can also add the increment in solid eggs such as called EXO increment by every step in while loop we have performed an activity that it is saying that in while loop. Let's say that while x is smaller than five, digital, eight and nine is equal to high let's just give it a delay of let's say 500 milliseconds Okay.

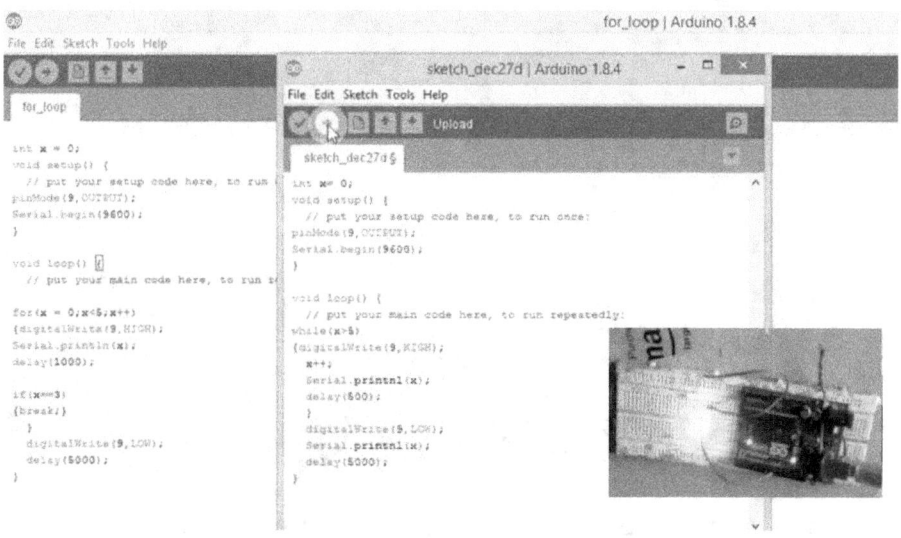

Now after exiting it should be given a statement or else it will again run the whole for a loop. Nine comma low it will be running for the next five seconds. To give an idea of 5000. In a while loop, if the x is greater if the X greater than five then it will exit the while loop. Until then It will be running continuous local adjusted. We can also print it on the screen. So let's just print it on the screen, right serial dot print In, it means printing next line, each value, let's say excellent. Now just copy the whole statement and you can also paste it here for a better understanding of the program. Okay. Now, I have stated that there should be a syntax error, you can see that the word does not turn to orange. The same error here let's just remove it and let's just upload it to the world. ledger save it as a name of while loop. It's uploading and uploaded. Now you can say that in the digital serial monitor I think there's somewhere in the code while x is Moeller I have said too that there is an error now changes the syntax uploaded once again. While x is smaller than five, then the light will glow after x is greater. After a while, x is smaller than five the light will grow after getting a value of x that is greater than five the light will with the light doesn't work the light will know to glow Thighs is the simple syntax of the while loop and we have performed it with the function of increment in it.

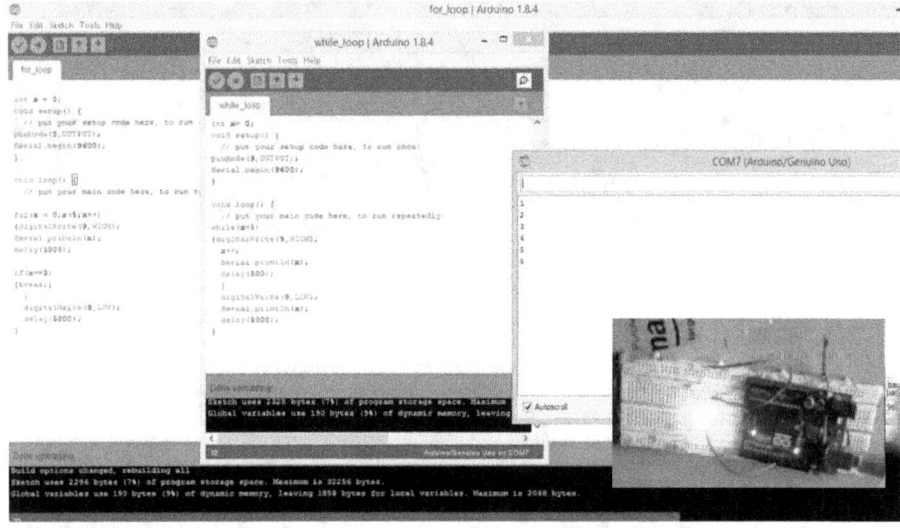

You can see that the loop is running continuously after the while loop it has been entered into the next lines, but you can see that the value of x is when the value of x has been declared and it cannot be changed. So the function will know to allow the LED to glow once again. Thighs were the example of the while loop and we have declared the ledgers recall the differences between them all for loop is used as a simple loop. For the declaration of a variable, it has a much greater extent of control over the loop. And it can also both the loop can perform the brake operation. Now, while loop can also run on the Boolean statements such as one is equal to one of the physical tools that are available in the world such as one is equal to one, or one is equal to one or you can say that two is equal to anything that can be a physical statement. And it can also run on the function of increment method you can see that we here perform the same function as the for loop using a while loop but there's a difference in y log the last value is always less than one. In the thighs case, you can see that when it is smaller than five. So the last week you will be reading here is four while if you are performing the same function in the for loop, the last value is We'll be reading is equal to five in thighs in while loop it will be for any info look it will be five.

USING ARRAYS

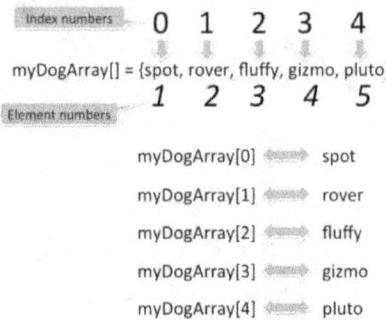

We are going to discuss arrays, well array can be said as arrays, arrays are the integers which can store a large number of values into it and they all have a specified place or a specified number to call them. Thighs method is used to store a large number of values and to call them when it is required. We have discussed previously the for loop well array can be easily called with the help of a for a loop. Now we are going to discuss the array and making a small program out of it. Okay, first of all, open Arduino. Make a new document, make a new open Arduino Make a new sketch now get here. First of all, we want to see the value. So we are first going to call the serial monitor values SPRI n sorry for the mistake. Serial dot begins 9600 with parenthesis Okay. Now, let us get to the topic or array can be called as an integer array, my array A no difference used to brackets. These brackets are thighs syntax which is used to say that thighs are not an integer but it is an array give a value equal to Now use two curly braces in these curly braces, you can place the values of the integer I say six commas eight commas three commas two commas Six again okay let's say nine commas 22 okay see there how many elements are there in array 6832 922 Now listen carefully that the first value of the array is not my era zero, but it is the value. Now listen carefully that in thighs array we have about six elements. so you cannot define the first value by using the method my array one for the first value the array starts with the values 04 if you want to get prints six, you should use the value my array 046 let's just try it.

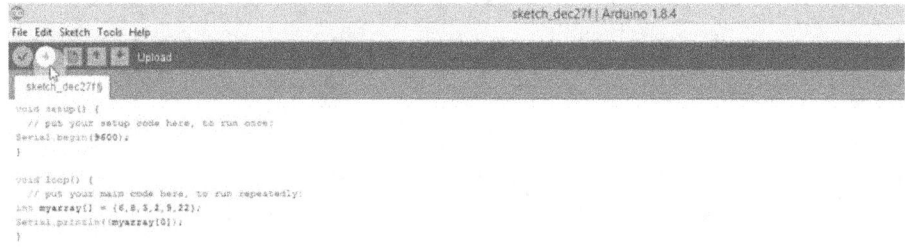

Say serial dot print ln say my array use the full syntax made to be decades and say well use zero. Okay, terminate the line uploaded to the board. Wait a second. I'm just plugging my Arduino. I'm uploading to the board. And let's just upload it. It shows that it is giving me an error.

Let's see. I'm turning the serial monitor off thighs Wait a second. I have not closed the bracket. That was the mistake. Okay, let's see now. Just see it's uploading it's compiling and wait a second. There was some problem before so again, let's just try it. It's been successfully uploaded legacy what was the value zero. And it's six.

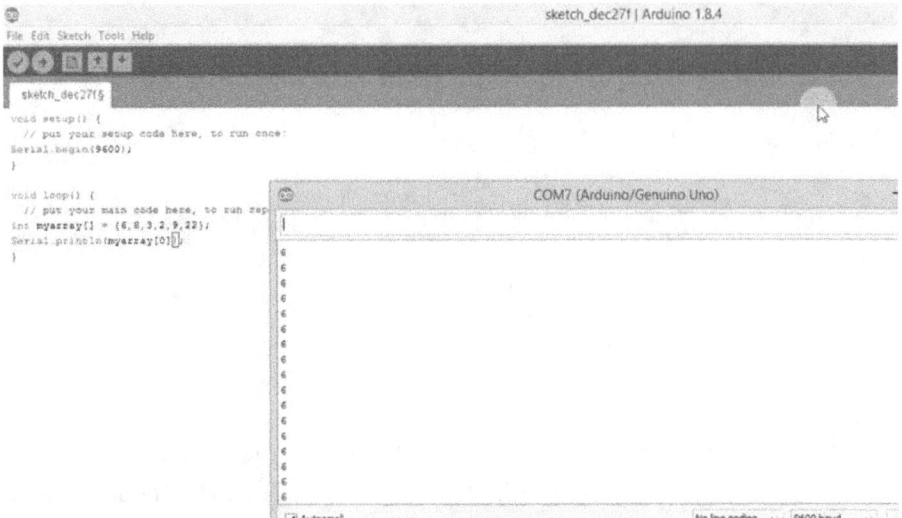

If you want to call the next value, just add it from zero to one. Now uploaded once again. You can see the value of eight is getting on the screen. Thighs are because when we are calling array, my array element one It is saying that it is the second in the line.

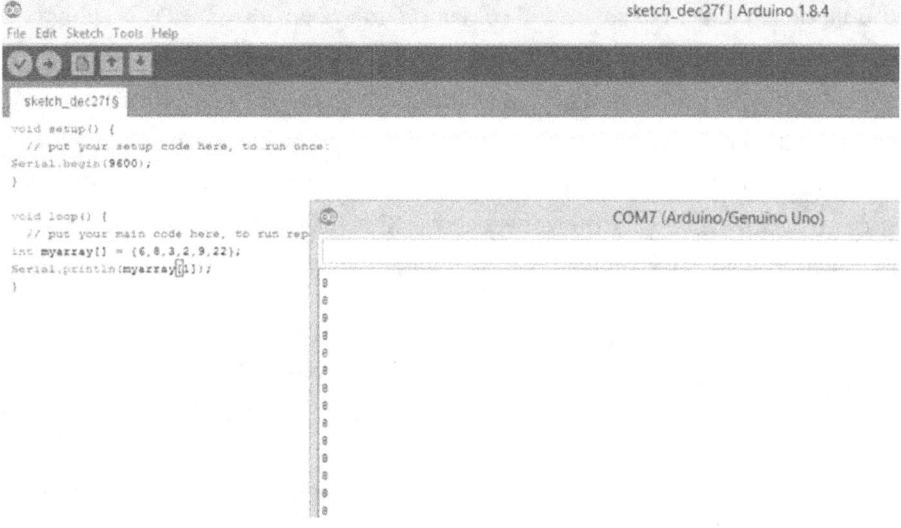

So, for a second you use one for third use second and for first you use zero thighs is the method for calling an array, but what about calling the list complete list of an array is it possible? Yes, it is possible with the for a loop as we have discussed earlier.

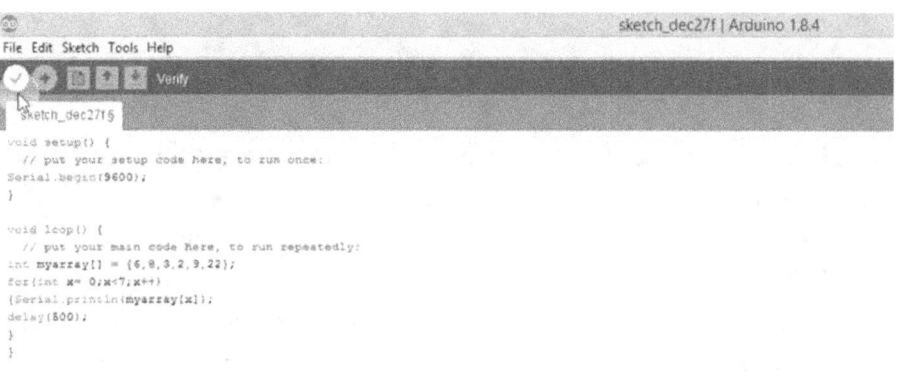

Okay, let's just try it. For silver possible delete everything about nothing more. right for you two parentheses, x is equal to zero x is not defined earlier. So we have to declare it I nt integer x two is equal to zero give a semicolon x is smaller than seven because there are six elements which should be smaller than seven as I told you earlier. x increment Okay, does the syntax is correct. Let's just do two curly braces. CJ load printer Print In I am calling my array a llama array call into brackets as is the method and what should be the value here the value should be placed here is x, but there is one problem because of the speed of the processor. We have to make it slow so that we can view the values clearly Okay. Give it a delay of say 500 milliseconds Okay, terminate the line, compile it. No compilation error.

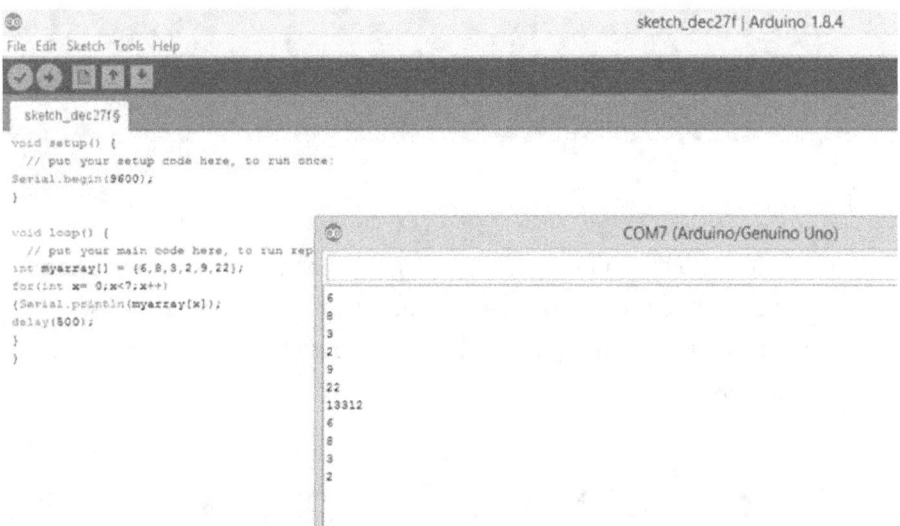

So upload it to the board. We have loaded the word successfully. Let her do it once again. Okay, the program has been uploaded on the board. Let's see in the serial monitor. Started once again for veterans Standing first element 6832 922 every element is being called there is a systematic error that we have used the word seven so, it is also taking the seven which is not possible. So, it is giving you garbage when you come into the program right it will lose six it will not give me an error okay give it a value six again uploaded and call the serial monitor now you can see that without an error the program will run.

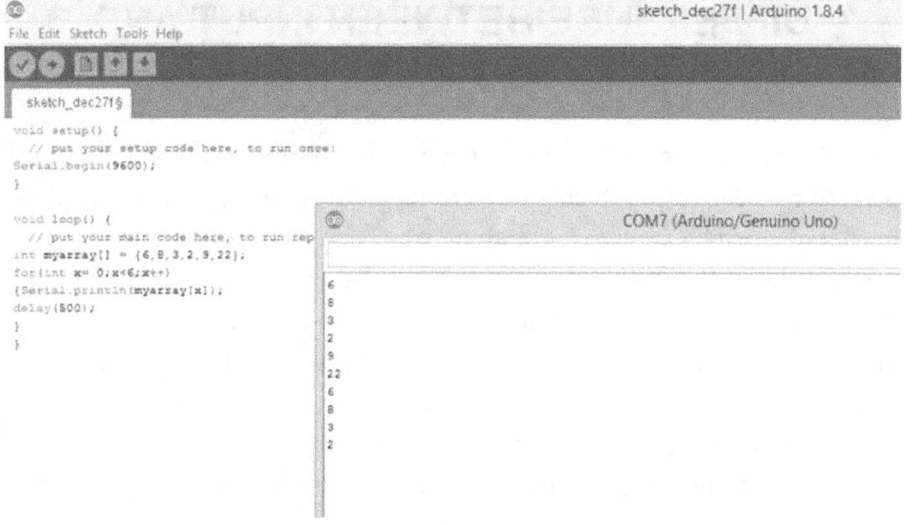

okay, there is no render program and the array is called calling all the elements as required. But what is the use of array the array is used in most of the cases for calling multiple sensors or multiple you can say LED lights, for example, L is an array of LED light can be called by thighs method so that multiple lights can get on and off at a specified time for as per the for loop or whatever your condition is you can connect Roll it completely and you can define 563 any PIN randomly and you can call them with the help of array?

DIFFERENCE BETWEEN INPUT AND OUTPUT

First of all the difference between input and output, the thigh is a theoretical part. So we have to get, thighs is a theoretical part so we have to get through it. First of all, there are two methods only by which the pins of an Arduino can get access. First one is to give a digital input or the second one is to give it an output. The output and input can be both digital or analogue as per the pin description. Most of the pins are capable of giving an analogue output or input while every pin in the Arduino board that has a number is capable of giving input and output in the form of data such as the analogue pin a zero to a five will be turned into a digital pin by using the number Dale digital pin 14 to detail pin 19 Okay, first of all, the input in the pin is given in the form Have a voltage if we connect the data pin to read something and connected to the ground then it will be given a value of zero while if we give a voltage of five volts which is connected to the world in the form of the ground then it will be given a value of the same applies for the analogue while there's a difference between digital and analogue is there an analogue you can get a measurement value. For example, if the voltage is three then it will be given a value of 600. For the voltage of five, it will be given a value something about 488 40 anything similar to that it will be a range of value from zero to one zero to three while reading and for writing it will be a good way to value humans to say that zero to two days to it minus one that is equal to zero to 255 will be the read writing value, while the input value for reading is zero to 1023. Thighs are because the analogue to digital converter is configured in such a way in the microcontroller that it can read tours to 10 values as input values. to eight will use will be given as output for the voltage sources, but the scale for it is same it is zero to five voltage only the difference is that in reading you can get a four-time precise as value as in writing Okay.

Now let's discuss the code and what is the difference in it. Here you can see that both of the forms will have the same method for input or output such as First of all we are going to discuss detail in what is the difference in codes while writing a data input or output. Both of the methods only have one difference between a mode for input it is written capital input, and for output, it is returned capital-output in the same method, but in when it comes to program in world loop there is some method difference is available, they submit but when it comes to word Look, there is some difference for calling the pins for digital right from And it is very simple just declare the pin as digital right give the PIN and

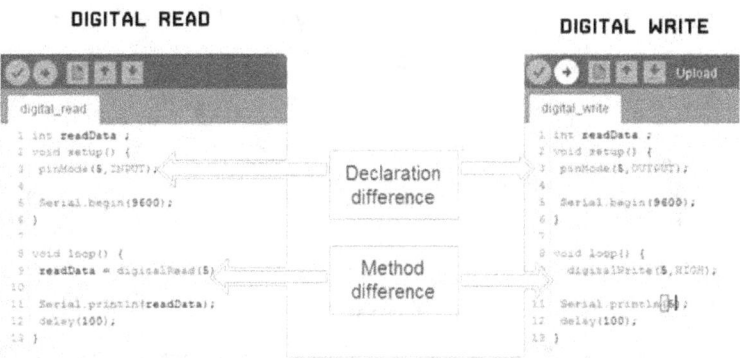

set the value for digital output it should be a low or high while when we come to digital read something, first of all, we have to declare a variable and then variable should be storing the value of the digital read and the pin there is no declaration of lower high because it is the read function it does not require a pre-built value it will be reading from the environment or from the open where it is connected. Okay and the same method you can see the print value using the same serial dot print In for digital read as well as serial dot print In for digital write the method.

MULTIPLE LED

We can run them independently while controlling about three LCDs at the same time, okay first of all about the LED there are simple light-emitting diodes that are connected to Arduino when the detailed pin is high, they get glued in when the pain is low, they will not blow okay thighs is the theory of muscle okay thighs is the theory of LED. Now what we what would happen if we want to control a large number of LED say that they are connected to PIN five, nine and three, but we cannot control them randomly. If there is no method if we are giving if we are going to write a program for that it will cause a lot of light at the program will be unstable. So there is a method by using array as we have studied earlier. If we use an array to control a list of you can control a list or an array or you can control a list or an array of LED arrays together with Arduino board and some small programming method that is controlling led with array Thighs is an important part of physical computing so please concentrating let's just begin with opening Arduino. Okay, first of all in the setup you can see that I have connected the LED to PIN two, PIN three and PIN four, okay, that's the pin of LED so first of all let me make it simpler okay we are in volt setup. First of all, we are going to call the pin mode of all led so that we are getting to we are giving them values High and we are getting the output we are setting them as output we are setting the ladies as output okay calling three-pin modes to give it a two give it a three and give it as four now capital-output or you TPU ti minus mistake output now just copied and pasted on all three sections okay. The first function void setup has been completed. Now we are coming to the main point, first of all, making an array I am giving a name into l e d s int LED is given do brackets that is enough given to equal to do two curly brackets for calling the array. Thighs are the place where the value of the arrays is stored. Okay, we are given PIN two, PIN three and PIN four here okay terminate the line. Now, it is simply called a for a loop. Let me show you an example. Let's just say D ID it I digital WRI te digital right LVDS l e d s, give two brackets and let's just say one. So the PIN three LED will glow. Okay, let's just take the program and we are compiling it there is some error and the error is it is saying that let me correct the program.

It is a simple mistake over dinner. Okay, let's just do it. Once again. Okay, everyone It was my mistake I did not do the argument the value that the LED should be placed higher low and giving it value by now you can see that the LED number one will be glow.

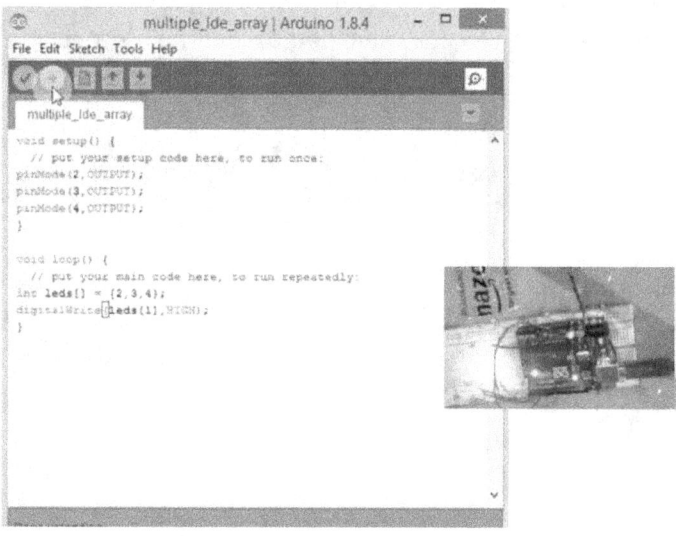

Let's just see it. Okay, you can see that the LED that is connected to PIN three will flow. But what about PIN four if you want to test it, you can do it. Just go there and say So on the array through that means the third one, the third one, you're connected led number four. Okay, let's just see it. Okay, you can see that the fourth LED is glowing. But what about if you want to call them in a sequence and give us a simple example over and give a simple example of a program for working it in a continuous model? Okay, let's just try it. Delete the line. You can use a for our for a loop. Do parentheses, say x is equal to zero, as we do in for loop, give you a semicolon, say x is x is smaller than, say three. X is smaller than three. Do parentheses, x plus. Okay? Now do two curly brackets. It's simple. Just write delete. All right. Digital wright l e d s le days do two brackets. Be careful about writing the bracket I have done a lot of mistakes here and the value that we are doing it is x we are giving x and the statement is saying that it should be high okay terminate the line. Now again thighs are something you should be careful once the LED are blown they cannot be turned off your to mention that to turn off the LED Okay, come here, copy the line after turn after the finishing of the loom, write three lines and them in all three lines we are saying that the value of LED should Okay, let's do it. Taking a second just wait for it. And the height should we return as low because we want to pause we want to set them at low speed for a second. Okay, let's just do it but for how much time it will be done in within a fraction of a millisecond. So we have to delay for let's say a second we are giving it a value of thousand-millisecond delay. Okay.

```
void setup() {
  // put your setup code here, to run once:
  pinMode(2, OUTPUT);
  pinMode(3, OUTPUT);
  pinMode(4, OUTPUT);
}

void loop() {
  // put your main code here, to run repeatedly:
  int leds[] = {2,3,4};

  for(x= 0;x<3;x++)
  {digitalWrite(leds[x],HIGH);
  }

  digitalWrite(2,LOW);
  digitalWrite(3,LOW);
  digitalWrite(3,LOW);
  delay(1000);
}
```

Let me explain it. So, first of all, we have called the pin mode have led to three-four outputs. Now in world blue, we have declared an array that is 234 that is the values we are calling here. For the next line, we have made a for loop which is saying that the LED x is will be high for let's say a second let me correct it. I will be blinking it for a second 1000 when the for loop is led array value will be called and it will be gone and it will be bright for one second and once also I've read the program will end and it will set all the ladies to Lola just take now let's just try it once again that is a mistake in the program I said led three times so let me correct it and now let's see 123 stop all the ladies are blown as they are. So if you want to see it in the serial port, it can also be done here Just let me call the serial monitor is e r l l dot begin. We can use different values of begin, such as 115200 but there were higher baud rates and we will be discussing it later. In the line, you can just call it serial dot print ln the value of x. The value of let's say x. Now we are not using the value we are using the value of Ra that the X is selected for the second Okay, again

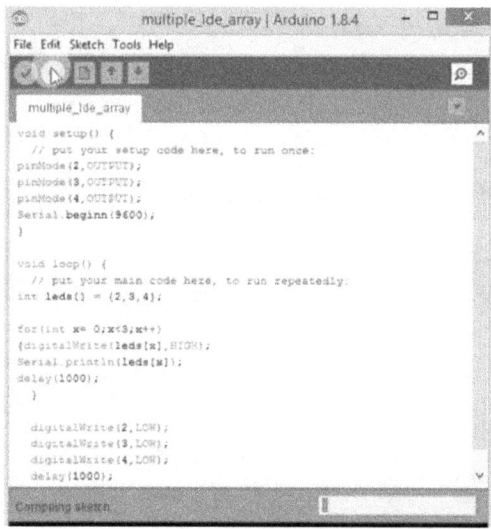

uploaded to the board there is some correction required serial dot begin there are two ends corrected uploaded to the board. And once the upload is completed ledgers open a serial monitor the program will restart and Sr 234 shut down on second 234 and again it will be shut down okay by thighs method

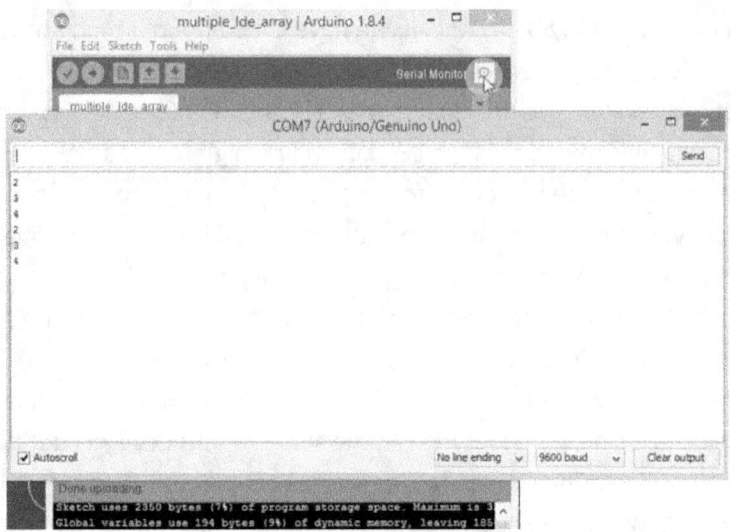

we can connect and we can control a list of digital pins that are connected to another value randomly it can be any pin such as 459 etc. It can also be used to control analogue pins and it is very useful when that will containing a list of sensors and motors.

BUTTON INPUT AND LED

We are going to discuss an input such as button and giving it an output such as led Lu whenever we press the button to state high the LED will glow in whenever we release it the lady will go back to offset okay now, first of all, I want to tell you that thighs are the most basic type of program that is being done in the physical computing it's just a simple circuit with the help of an on the button you will get the LED on and off okay. But the important thing is that in thighs in thighs program we are using the logic of the microcontroller you can say that when a microcontroller sets one pin high it will perform a logical operation then it will send the output of the digital pin that is connected to lead to high so thighs are performed SM logical operation for the LED to great pride. brightening, okay. First of all, let's just try to write the program.

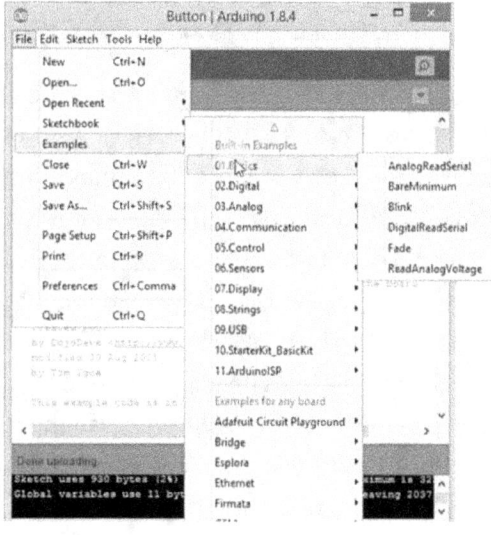

The program is available in Arduino we example in Section detail named as a button but we want to make our program our program, but we want to make our program so we are doing it from scratch. Okay, open a new tab in Arduino. I write the pin modes. In the thighs case, we need two pins one for input and one for output. Okay, let's just do it. Copy the syntax Just a second.

We are copying the pin mode. Okay. We have given two-pin modes say led to that is input it means that it is taking the signal from it is reading the signal For state high or low okay and we are selecting led pin three as the output or the pin at which the LED is collected Okay that's it.

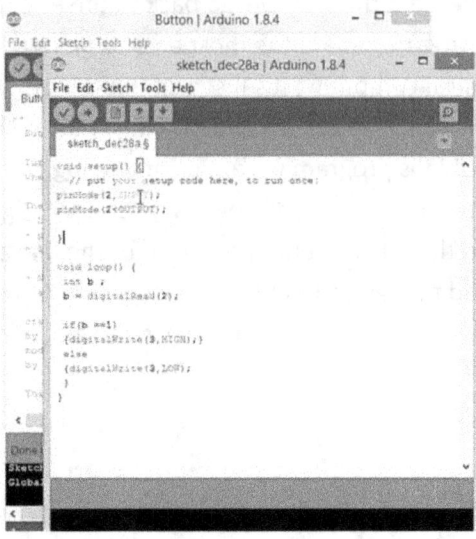

Now here let us declare an integer say int intp before what n I n t v okay we have declared an integer called v. Let's just give a simple example for right reading the details. We need v to store the value of the reading that is done when the digital pin two is getting high or low. So it is simple writer di dip l capital R e A. D digital didn't turn into orange letter P number two in our case it is PIN two Okay, so the LED is the LED pin. So the integer we will store the value of the digital print to in it. Okay and just write a simple logic fob is equal to one we can say big it LWR p digital right p number three comma H I g h high or we also need an assistant when it Elsa the IDI to WR IPA digital writer you tell write a PIN three l o w okay let's just see the syntax of it. At first, we have declared two pins such as PIN two as input and we are writing PIN three as an output the PIN two will be used for reading the value and PIN three will be the PIN two will be used to reading the value while the PIN three is connected to an LED to us and output. Okay and in Here we have declared deserve what for the better for the betterment of the practice we are always trying to declare the integer at the top of word setup Okay, we have made it here okay it is our V and let's just say the initial value of integer v zero it will be a full betterment of the program and improving the performance Okay, the first value of integer we will be zero but after entering in the void loop that is or B, we will read the value of the V number two as digital AP number two goes high, we will be one and if the PINs two goes low the V will be zero. Now, for the if statement the condition is a B is equal to one or we can say high then the operation of the retail Right Thing number three will go high or it will blow the LED and then it for the asset when we must be giving an S statement saying that if the value of b is not equal to one then light will turn off or you can say the number three will below. Okay, let's just compile it.

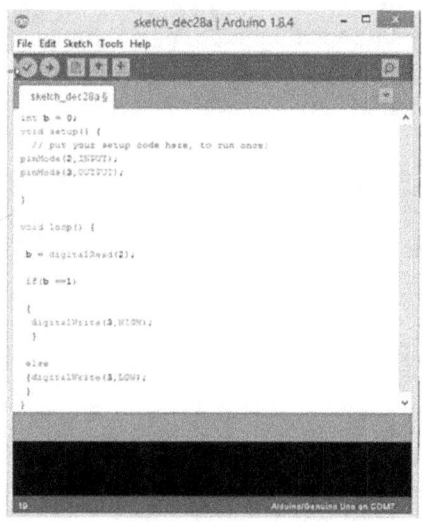
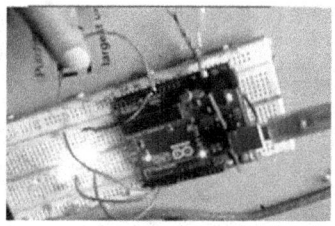
HIGH

LOW
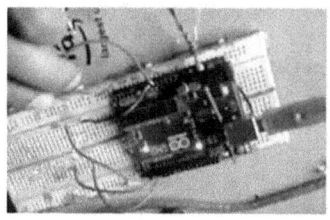

We are compiling and giving a name we used to in button okay it's compiling Just wait a second. Okay we are uploading it to the world and the upload is done okay, let's just take it. Here you can see a wire is connected to five volts and thighs are PIN two. Whenever we connect thighs wire it will turn on I but remember one thing that because of the garbage will use led in turn high for a few more seconds for that we have to connect the high PIN to wire to ground to get it low Okay. See, you turn it high. To turn it low, you turn it high, you turn it low, it is better to use wires in place of a button for testing because in button you can get the garbage value or you can say the micro values are initially getting high and low causing the transient which will give a random value of high and low which will make the LED to flicker Thighs is because the processing speed of the microcontroller because it is in milliseconds so immune the time removing the transient we are directly connecting wires. Let's see once again thighs is because of garbage value now SR turns on and so, on terms of okay thighs is the use of button value and button will be very useful for many applications because in most of the cases the there is no button but there will be sensor there is digital sensor they will be given a value of high or low and for that system of giving getting value high or low you need a simple logical program such as return here to getting the value and perform a retail operation for and perform an operation such as detail right in high you can be using another motor or anything but in our cases, we have made it the most simpler format such as growing an LED okay the code will be presented withighn the beginning of the PBT.

INPUTS BUTTONS

Arduino and what I did about what in just last time I will explain how to connect Washington to Arduino to control the requirements are computer on a laptop Arduino microcontroller. You speak to city and Adapter. If your microcontroller does not have us Vibart appropriate to us via cable. But I'm not going to switch minutes and need connecting to switches push buttons and momentary contact switches so that we know is a straight forward. I wish what Don is a simple device that can lead to circuit one end of the button is connected to ours usually a low voltage in which 5 volt and Arduino is ideal.

The other is connected to how did you Dalman when the switch is flipped pressed are toggled the circuit. Is that open or closed. The digital bin simply returns if there is a 5 volt. Or is it over. Now in the following similar program that can show you published what it is used to control the lead. Is it us. First we start by initializing the global variables event button been equals seven semi-colon where the lead of the Bush button is connected. Do you know how you use your number 7.

```
int buttonPin=7;
int ledPin=13;
int status=0;

void setup() {
 pinMode(ledPin,OUTPUT);

pinMode(buttonPin, INPUT);
}

void loop() {
  status = digitalRead(buttonPin);
  digitalWrite(ledPin, status);

}
```

The end is short for integer. The fact that the be in the bottom pan is capitalized. Makes it easier to see that it's actually two words. Since the space cannot be used also the line must be ended with a semi-colon as shown and then equals 13 semicolon where the lead of the lead is connected. Here I have chosen the best and lead status equals zero. Sénéchal on I Givat initial value of zero in which to be off he under the void. We write mode been out but don't realize that Ben has an output then been mod. Putting pan and bought to initialize. Published on Ben as unbooked he and Altovise. We this step. Thus he Chua's digital read button pen to McCollum which is used to read the start of the pushbutton and to start it to status. Then the line digital right. Ben status semicolon so that since yesterday. Us two that had been if it's high the lid been turn on and if it's law then it will turn off depending on the status variable. And now you are good to go. Just connect you are doing board using USV cable and no luck on. Like I showed you in my previous lessons. Thanks for watching this lesson. If you have any questions please ask it. But you and I are wired to Unix on.

ANALOG INPUT IR SENSOR

 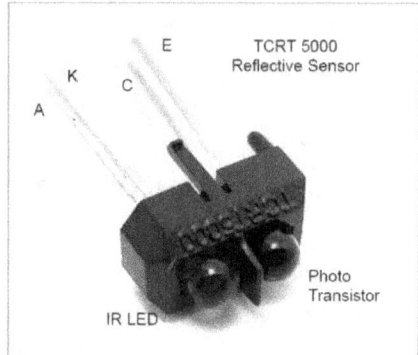

I'll probably just be careful and I just don't censor in front distance and so on are useful for measuring distance without actually touching a surface to three wires protruding from a distance and so isn't that far evolved. In most cases ground signal these are almost always color coded with black as a ground red as a vault and white as the signal. If you're in a for distances so did not come with me why are you and I need to find.

```
Int x;
void setup() {
  // put your setup code here, to run once:
  Serial.begin(9600)   // initialize the serial communiction at bude rate of 9600
}


void loop() {
  // put your main code here, to run repeatedly:
  x= analogRead(A0)
  Serial.println(x);

}
```

I'm probably a connector or so that wires directly to the leads ensure the bends Soledad. No not contact with one. So you can attach wires Nick that it could have evolved on that do you know connect the black wire to ground on the Arduino. I have been on that do you know in this case which is it like Shawn in the picture. Since the sensor is connected to the game board of that we know the code will be like this. We first start by initializing the variables X semicolon which is initialized integer variable x then write it up we write C and begin nine thousand six hundred semitone which initialize the serial communication at World Trade of nine thousand six hundred. Under the why do we write X was agreed in between the brackets number seven semicolon which should be data value from the same store that is connected. Place as usual then we're right on the brink. And in between brackets X then semicolon which the Sublette output on the screen are not dissimilar to the board and a change to the minatory as you move the front of the distance and so on closer to and away from a solid object. What are the values should the change between 0 to 1000. 23.

ANALOG INPUT POTENTIOMETER

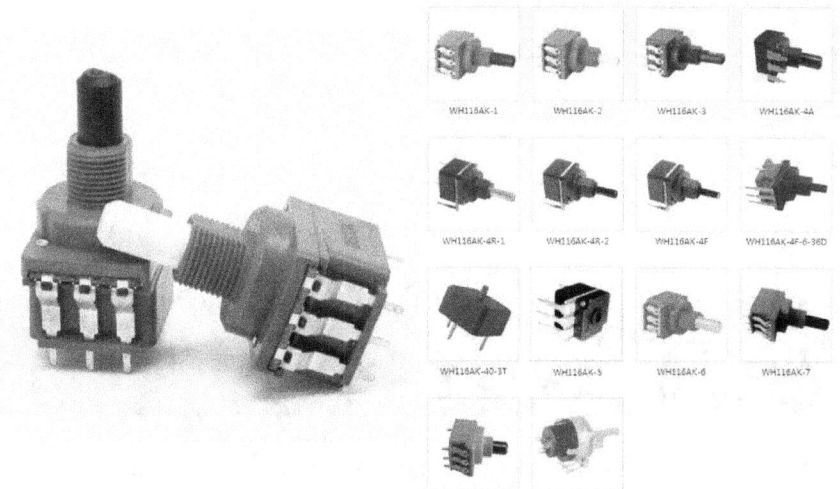

You asked me to say adapter if your microcontroller does not have you as we want everybody to be cable show me to minutes and then these are the types of meta the circuit of this goes on should look like this. Connect the meter Tobin's 0 5 volt and ground them indeed. Is there one took a nick to that bin and the voltage via's on the spin orientation of the other two bands. Does not matter. Now I will show you and explain to you every line of the code which is required to precede this lesson we will start by initializing variables as defined.

```
int sensorPin=A0;
int sensorValue=0;
int ledPin=13;

void setup() {
  // put your setup code here, to run once:
pinMode(ledPin,OUTPUT);
}

void loop() {
  // put your main code here, to run repeatedly:
sensorValue=analogRead(sensorPin);
digitalWrite(ledPin, HIGH);
  delay(sensorValue);
  digitalWrite(ledPin, LOW);
  delay(sensorValue);
}
```

First and since I've been he was easy semicon then and since our man was chosen only to describe what value I wanted to present then to just sort of man is equal to zero where is zero is not zero on the Arduino on it's on. A zero is not a reserve term. However when used in context the system recognize it as an audible been zero. The line must end with a semicolon. Like I said before by declaring a variable this tab you can use the term which in this case is some sort of pattern throughout the code. And instead of as you will there are two main benefits to this one. It makes the code more descriptive too. If you want to change the value of the variable you only need to do it in one place then.

$$3/5 = x/1024$$

$$x = (3*1024)/5$$

$$x = \sim 614$$

And since value equals zero semicolon the term since value is not reserved term either then lead equals 13 sénéchal. Once again the term Libin is not a reserved word in Arduino. It was chosen to describe which bin was connected to then the value of 13 is a normal value. But just like a zero when used in context give reason spent 13 however you will need the following code and other void to been made an output which is used to declare that the lead has an output at last we read the void which will be repeated indefinitely since around you it was agreed since I've been cynical on this line use of the term and I love read in order to read the voltage of AM and I know been most microcontrollers used them. And I look to digital conversion in which voltage conversion which is due to the dense power of a number equals 1024 therefore voltage of zero corresponds to a nominal value of zero voltage of five corresponds to an added value of 1024. Therefore I value everything evolved well correspond to an omni value which can be calculated like this.

```
int sensorPin=A0;
int sensorValue=0;

void setup() {
  // put your setup code here, to run once:
Serial.begin(9600)  // initialize the serial communiction at bude rate of 9600
}

}

void loop() {
  // put your main code here, to run repeatedly:
Serial.println(sensorValue);

}
```

The sketch name had to be modified. Sketch names can only consist of ASCII characters and numbers and be less than 64 characters long.

I wrote that equation. You can get the inside doing this three divided by five equals X divided by 1024. We were X on one side. So we tried three multiplied by 1024 all divided by five. That answer will be approximately six hundred and fourteen. Then we died. Did you get it right. Hi. Which Dan's been on delay. The delay should take the delay time from sensor value which stops the program for how many seconds of this interval you then Digital's right. Did Ben know which devastated ben of the delay had been taken from sensor value. It starts with the program for milliseconds of the sensor value. As I said once the button meter is connected. I bought this sketch to the board and it changed to the serial minute as you read said the knob or slide the slider. The volume should change between 0 2 1023. Correspondingly valid will blink with a faster or shorter delay. You can now read the values and use them with your account then your function used here is and I disagreed. Where have been selected has been number 0 if you use it. Then Number five you should change the code to read and since around 5 if the system does not track check the syntax and ensure that God Blonsky get the next check the connections to the potentiometer on shooting that then indeed goes to the correct bin and other been aborted at zero volt and 5 volt. If you bought very cheap on all that and she meter has a chance it may be mechanically defective. You can test this using a mustimeter and extends to that model been out. I've been sit multi-meter to read Jesus stands up. Resistance changes slowly but is working. That is just dumb is idiotic. You need and you couldn't Shamita. Now what if you want to see the value yourself. Take a look at the code that I display here and that is as a new sketch before initialize it a variable instance or append equals zero which select then but then for the button Shimit to then in tenser value equals zero which is the variable to store the value coming from the sensor. But some of your functions will be used now under the void. The void set up under the void tab we write See look began in between the brackets 9600 semicolon which initialize the CNN communication at Aldred or nine thousand six hundred. Then under the void log we write C and D to print and in between the brackets since her value then the semicolon. This Since the value contained in the variable since value c via the US be blank and digital been run free by then and I'm not discouraged to Arduino. Once it's done based on going in fighting glass located towards the top right of the window this is silly I wanted to add monitors communications begin sending and receiving by the Arduino here. You must verify the Godhead is also nine thousand six hundred on. You will see garbage. Thanks for watching this lesson. If you have any question please ask it in the queue and airborne. See you next lesson.

ANALOGUE INPUT WITH EXAMPLE

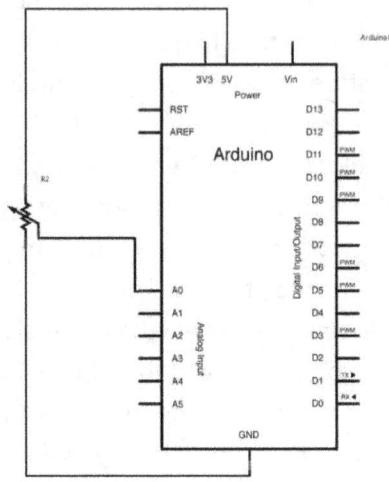

Analogue input is the method that is used with the help of a potentiometer to show the voltage through the board and says that the word can display the voltage on the serial monitor, but here we are going to study that you can write manually analogue values into the program and it will be giving us output to the LED brightness so that if you are given the value 255 the LLC will grow to its maximum but if we are given the value of 150 then the LED will grow up to 50 percentage. And if you are given the value of analogue right 50 then the LED will glow very deep. Okay, let's just try it with an elven example. Okay, first of all, open the Arduino. Let's start with writing our sketch, we want only one value such as led pin which is connected to PIN three. So we said PIN three as an output. Remember one thing that should be very clear that analogue output can be generated only on the pins that have a special type of symbol on it. Well, you can see here the pins that are connected to the analogue have a special type of symbol on it which is named as PW because analogue writing in Arduino is done with the help of pulse width modulation. Okay, let's just write the program. First of all, begin with the mode. We are doing a pin mode or for PIN three as our Output okay number two set us output okay we want an integer that we will be storing the value for analogue we said is integer v is equal to zero its initial value is zero okay come here right I empty no there is no declaration for your V is equal to less than 100 v given the value for integer such as hundred. Now write in L o d analogue capital W RMIT analogue right thighs will be calling the analogue PIN three and giving it a value that is stored in the variable we see the syntax declaring an integer calling the penis output giving the integer value and writing it as an analogue format on PIN three with a value of integer v.

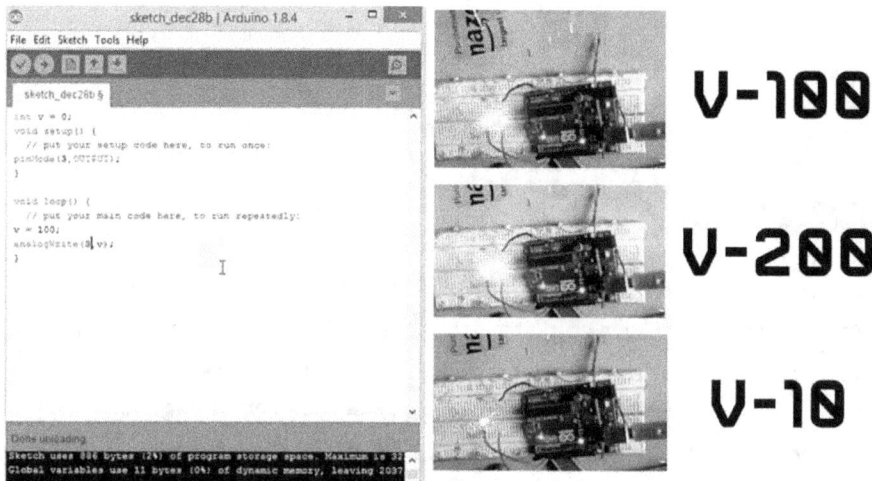

Okay, let's just upload it to the board and see here. You can see there the light is growing but if you want if we give you the value From 100 to 200 just see the brightness of the led turns more bright again route 100 and see here the LED glows dim, but if you want to see real performance started at 10 now see the value of the brightness of the LED it's glowing very low. So, if I thighs method we can directly give output to an analogue pin to perform the operation which has happened which happened because by writing the variable to the analogue right we are changing the duty cycle of the builder, which in turn from digital to analogue converter, which is a bit in ATMEGA three to it that is used in Arduino Uno will give a duty cycle which can be said as 10 divided by 255.

That will be the duty cycle. Again Mega 255 they will the highest point of rightness. See the lead is going to it right. So if we want to see how the value is explored or how much is returned in the Le interest, how much value has been stored in the variable you can see here let's say 9600 9600 here and come down. Write a value sedar dot print ln. We are storing the value in V so we are writing here okay uploaded it is brightening, okay. See ya. The value is 255 that is given by analogue well now save you 25 The lady is growing at a very tame pace and the value of the analogue rate is 25. And if you want to find the duty cycle just do a simple division of 25 by 255 which is about 10% duty cycle okay thighs is the method for finding the duty cycle and by thighs method, we can control the analogue right or you can say the analogue output with the help of simple programming.

LIQUID CRYSTAL DISPLAY

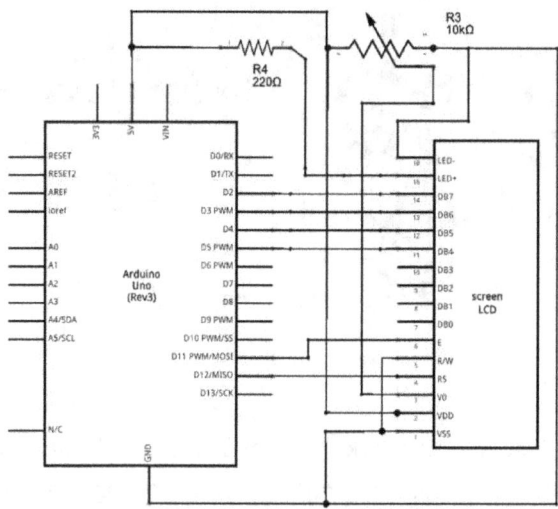

Well, LCD is an important component in Arduino because it is the main thing That can give you the display of the program or you can say the variables that are inside the Arduino a standard on so it does not require a computer, such as in serial monitor whenever you want to give a display, we need a computer while using thighs LCD here. You can get all the displays on the LCD so that we can make Arduino and complete the system in a box and can place it anywhere without the help of a computer and we can see the variables inside it can use to measure temperature, moisture and you can do your custom prints on the LCD.

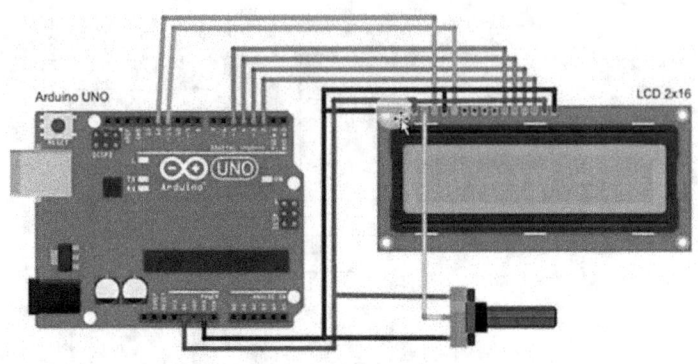

Okay, now let's just see the block diagram of the LCD. Well here is the schematic of the LCD remember, important things Thighs means are for the light thighs means are for the power. In here thighs important penis for analogue display. If you convert the ad to rotate the dial, it will give you a value that will make the LCD go dark or dim. Okay, I will show you Just as you can see that with the help of rotating with a dial the potentiometer of the potentiometer you can see there the display of the LCD changes and which gives us a dark or allied view. Okay, here the schematic these are the four basics were which are used to come, come which here are the four basic software that is used to convert the whole the embedded program written in the Arduino to the display function of the LCD. Okay, now let's begin to write a simple example. First of all, remember one thing that in Arduino there is all so a built-in library for LCD and there is a simple program of LCD also written in the Arduino.

We are going to see here the program. First of all, we need to call the liquid crystal library so we are going to call it with the word liquid cristal.edu giving the close bracket now we are need to we need to declare liquid crystal element we are declaring liquid crystal element LCD and we are giving the PINs as PIN well which is an enable pin a PIN 11 which is for the display and PIN 2345 for the variables that are going to print on the screen. In word setup we are going to call serial lcd.to to begin in Word setup we are going to call the function as serial dot to begin it is setting the pin to 16 is to do it means that 16 columns and two rows are activated in the LCD, and in the next line we are going to print hello world. Thighs are the word that we can directly print on the LCD but just by calling the word into the brackets in Word Lu.

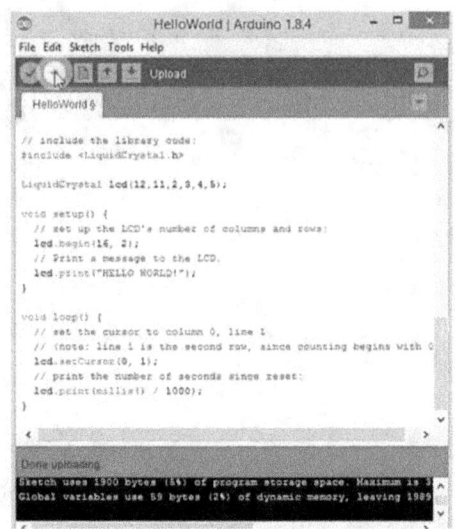

We are going to set the cursor to PINs zero and one that is giving the lower column here you can see thighs lower column the First section will be getting the variable point and here we are going to write a variable changing variable that is millisecond and we are going to print it okay let's just upload it to the world. It's uploading wait for a second and just wait for a second and here you can see that HelloWorld is written on the screen and it is counting the time. Let me make it clear display for you. Hello world and it is counting the second time in milliseconds by thighs method, we can connect the LCD to the Arduino and get our own desired display. Okay if you have any questions, okay, but whatever writing something is okay, we are going to write something Is acid come down here. And as we do coming LCD dot render, the writer says something Ah, ie we are going to write the value in a second line and the first word okay first section we uploaded successfully on the word. Now let's see where we can see the high is written here.

Thighs were the feature of the program here. But what if you want to read something else in another line, give the value for an hour write something else again. Let's just try with h IE high Please clear the space. Okay, we are going to print it in another line wait for a second it is coming and we are going to reset the Arduino. It will take a second and see we are calling my mistake. I have written it in the fourth row and we are going to write in the first row and the fourth column. Okay, we are going to upgrade the program and wait for a second. Now you can see that the word high has been more four columns to the left.

Okay by thighs method we can write the words on LCD we can print the variables and we can get real-time data on Arduino standalone without using a computer. Thighs are known as LCD and thighs is the LCD module I'm using here. It is manufactured by High tach and it is a very stable CD and it they are also the vendor A CD that comes in 16 crosses four they have some differences when diagram but basically they're working in sin and then require their custom library for running.

WIRELESS POWER INTRODUCTION AND HISTORY

So while this board came to remove these connections to be able to feel about wirelessly.145 So it means that transmission of energy through the air it's the process of transforming electric energy145 or Bhagwat over a distance without wires.145 Also known as thirsty and wireless power transmission power source which is that transmitter provides145 power to the device as well as capture devices the receiver captures it.145 Eliminates the use of cables for transmission of power.145 This is one of the most wanted advantages of wireless power.145 We don't need to use wires anymore.145 We want to get rid of firearms.145 You can see lots and lots of smartphones are competing to build or enable whams charging which will145 make the users or the end-user life much easier.145 So in this course, we will know how this is done sphere inside the phone or using your Arduino or my145 controller board transmission by induction thraw resonators.145 We will explain that buoying later on.145 But for now, this is what you need to know about wireless power.145 Now let's talk about some history on this for in 1899 said Nikola Tesla was a fly on this power transmission145 this law in Litan 200 lamps the distance of 40 kilometres but only 15 per cent efficiency was achieved.145 As it is in the active mode most of the power was wasted and has less efficiency.145 This was one of the very first wireless power transmission experiments the forgotten invention is reborn.145 In 2007 a group of engineers at MIT come or came up with the idea to use a resonant induction to transmit145 more wirelessly.145 Id bought a 60 watt light bulbs from two meters or seven feet at 40 per cent efficiency.145 As you can see this is one of the bulbs or light bulbs that they are lighting wirelessly as you can145 see.145 Here we have a coil.145 And here we have a coil.145 This was the first practical example of interesting.145 This is the globe that accomplishes this task.145 Now before ending this lecture Let's talk about the efficiency chart of footrests.145 We all said that almighty you guys reached 40 per cent efficiency while and Nikola Tesla achieved only145 15 per cent efficiency.145 So this is the efficiency chart of what wristy or way of electricity as you can see here and we have145 the efficiency.145 And here we have the distance.145 You will not notice that as the distance increases the efficiency decrease.145 So if in meter we have a 75 centimetre we will get around let's say 95 per cent efficiency but when we145 move on when we reach the 225 boinked or 225 centimetres we know that we can only get 40 or less than145 40 per cent efficiency as you can see here.145 The black box means that these are experimental results.145 While the red dots mean this was an experiment but it wasn't done over a wide range this box means that145 this is the theory.145 As you can see this line the blue line is the theoretical line as you can see the results are around145 it.145 As you can see the main point here is that as you increase the distance the afternoons will decrease146 significantly until it reaches zero.146 The highest distance that you can achieve

HOW WIRELESS POWER WORKS

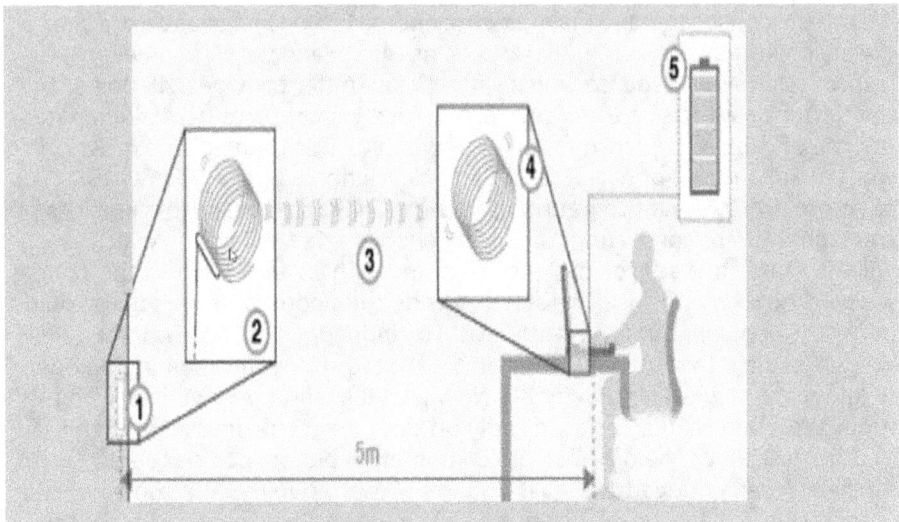

First you need to look at this diagram or this schematic which will help us understand how wireless power or works fears at this point. You can see a magnetic coil or antenna a that is housed in a box and can be set in a wall or ceiling. The second component here is antenna a board by means resonate at a specific frequency so this is a coil with a specific frequency on it. A resonator let's say that generate pulses. The fifth component here is the electromagnetic waves transmitted through the air from number to number four. As you can see we already said that this will run on a specific frequency so it will transfer waves not electromagnetic waves to the receiver here of the boring number for the fourth component in this there is a second magnet coil or antenna be fitted in laptop or TV. That is when its at same frequency as the first call here and absorb energy that was sent by the first coil. As you can see this is number five. It shows that energy charges the device so here we have a coil. And here we have a coil. This one since electromagnetic waves. This y z these magnetic waves and transfer all of them into energy or store them inside a battery or whatever are combined to have here. Now let's take a quick look at the basics. We have inductive coupling. We have magnetic field and we have resonance. These three components months must exist for the to be applied to any system. Now let's take a quick look at our system description. Now as you can see here on the left side we have ECAC means the AC DC converter. Now we converted the AC power of the electric current to DC current or direct current. After that we are converting the direct current to our amplifier or resonant frequency amplifier. Then we have impedance matching didn't work. These two items must exist to make sure that that transmitter and receiver are running on the same frequency and generating the same quality magnetic field. After that we have a source with a meter as you can see. This will generate magnetic power and send it to the device that originated here then it will go back to where I am on which will instead send it to the DC or to the R after DC rectifier. That is when that frequency to a DC rectifier will rectify magnetic waves or resonant frequency waves and turn them into DC or direct current which can be used by the Lord. This is a quick system description that shows you the internal component that you must use to generate power. As you can see or to Transfield electricity wirelessly if any of these components isn't clear for you I can explain it more just ask the on board now some things are. And let's take a quick look at this working principal. As you can see it's one symbol physic or physics wall at source. Electricity is the Transfield as you can see here into magnetic wave whine at the receiver device. The magnetic waves are Transfield into electricity. So here we have the wire. We transfer this power to electricity Tom McClintock waves and send it via air. Here we have a coil that receives these magnetic waves and turn them into a cannon that runs through all devices and turn them on. That's it. I hope that by now you got a clear idea of how wireless power works.

ADVANTAGES AND DISADVANTAGES

That ventures are safe fairly efficient and have a good range.148 It will help us get rid of a lot of cables so we will consume less kabar it's more secure.148 No risk of getting shot because it's wireless you can't touch a wire that is wrongly connected it can148 go to places where you cannot have physical cable.148 There is no one else which has the man advantage that we have for with resisting the need for the battery148 is eliminated.148 The e-waste is eliminated.148 Maintenance cost will be saved because there is no cable.148 It's just a transmitter undersea over the range of what can be increased.148 So these are the main advantages of what thirsty.148 Now let's see that its advantages Why bar transmission can be used only in a few minutes.148 For now, that efficiency is only about 40 per cent we can get more than 40 per cent efficiency in the way of148 less bio or electricity transmission.148 There is a great need for standardization and adaption.148 So a little overheating occurs because of different voltages of the transmitter and receiver.148 Refitting old equipment or purchasing new equipment could become very expensive because you might have148 a device that you want to turn it to a wireless power but it doesn't support that so you will have to148 buy shares in your equipment to transfer it to the wireless-enabled device.148 And this will be very expensive or you will have to buy a new device with a wireless power option.148 Which will be also expensive because you will have to buy a new device and there is a very good possibility148 of theft.148 While sport is just like any other wireless network that you have about it can come and steal your power.148 If you are transferring it over a wide range sure there are ways to protect it but its costly148 for now.148

RANGE OF WITRICITY AND WITRICITY APPLICATIONS

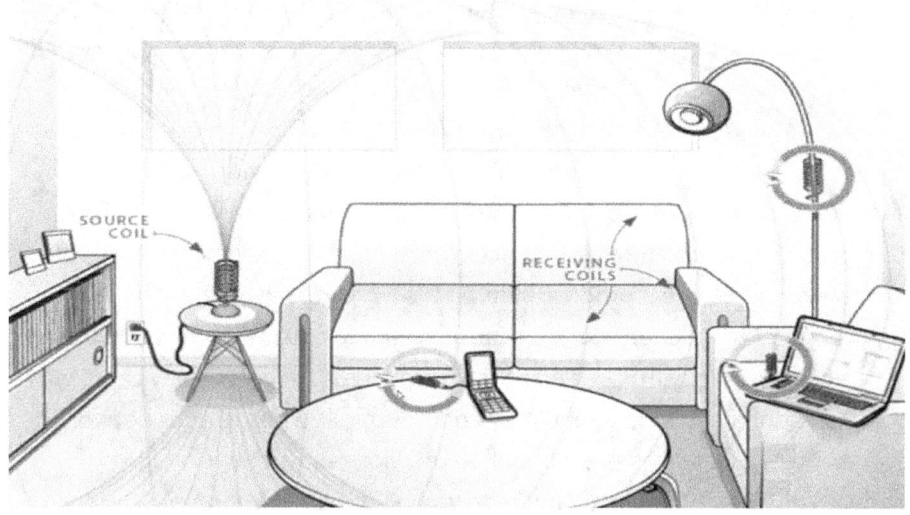

Let's start by talking about the range of futuristic wireless electricity can be for electric transmission from two centimetres to three to four meters depending on the equipment that you have. Scientists are working on extending this range and are expected to become 10 meters in no time. As you can see one source code is connected to AC power can transfer the energy for the electronic devices in this room. This is the whole barn with a silver coil. This is the laptop with Steve Alcorn. This is aligned with Steve Alcorn as you can see all of these receivers coils well received from that transmitting coin and will operate without wires. So this is an example of using wireless electricity transmission inside our own as you can see. It's about four meters wide room. Now as you can see these are three but batteries can be charged. It is just enough to place that device to tell one of the surfaces. This is a surface used when it's a lot of surfaces. You can simply place your device and it will be charged wirelessly without having to connect. Let's say a cable. You must be able to charge it. But it's a very small distance as you can see. You can charge it while spinning your hand. So this is a little of our sit back but they are working to solve it so that this system can be achieved with high efficiency. More than 40 per cent this is another example charging electric car using with writing as you can see this is an electric car has a battery and this battery is charged using as you can see here. Coils are connected to the AC boiler. They Transfield bore using magnetic waves. Current wireless technology is implemented in Bluetooth Wi-Fi and APS-C or near-field communication satellite communication cordless mouse keyboard mobiles microphone and headphones. These are examples that are using wireless technology in general. And here we are talking about wireless power transmission. Now, what does wireless power transmission makes products look like? It makes them more convenient no manual or charging or changing batteries. It eliminates unsightly unlikely and costly barcodes. It makes devices more reliable. It's never run out of battery power to use a product failure. It's by fixing the weakest link more environmentally friendly. Sure because it reduces the use of lots and lots of wires it reduced electric waste. So you don't have to throw away cable wires and you don't have to keep let's say adding copper wires to your network or more and more devices you can boil them wirelessly so it will help you reduce the amount of waste and help you help the environment.

TYPES OF MOTORS

We are going to learn about the motor and how to control the motor with the help of Arden, first of all, let's see the content or content is a type of motors. It can next content will be single DC motor, simple DC motor servo motor then after we are going to talk over stepper motor in the last three tutorials we are going to learn about the controlling weather of the motor and we are we will be experimenting on controlling the motors. Okay, let us begin to our first topic. Before starting type of motors I want to tell you that about a motor is an electromechanical instrument that takes electrical power as an input and gives out in the form of mechanical power it there are two types of the motor in the world the AC motor in DC motor in Arduino, we are mostly be talking about the DC motors of low voltages.

Okay, let's just start about the DC motors and their types first of all the DC motor it is a simple motor with two-wire you can see here it has only two wires. You if you want to change the direction just change the polarity one connected to positive and another negative and if you want to change the polarity just reverse the wiring and it will change direction okay thighs is a DC motor. Okay, next what is our stepper motor? Well, first of all, it will what is the servo motor.

A servo motor is a special type of motor that is used to give the values in degrees. It is not made to rotate continuously in 360 While it is meant to rotate for a specified degree, if you set the motor to run at 30 degrees it will get locked itself or 30 degrees, it will log itself to a 30 degree and when you try to change the direction of the motor if the program is running, then it will set itself to such a degree. Thighs are a very important motor for making toy aeroplanes in red because thighs are the very important motor for making toy aeroplanes because for the rudders if you want to set the value for 30 degrees and that will be causing another effect onto suppresses down while the motor will turn itself back to 30 D so that it can move to the urban Verizon The third type of motor we're talking is here is the stepper motor, the stepper motor.

The third type of motor we are going to talk here is stepper motors. stepper motors are of two types they are unipolar motor or micrometre. They are specially designed motor that can run on computer commands such as the row motor but they are used for more than 360 degrees run they are useful long run especially in CNC and 3d printers they can be very high-speed motors they can be very low or very basic motor they have a simple running mechanism which is set as a coil inside there is a magnet the magnet is rotated with the changing value of the direction of each coil which causes the motion in East magnetic coil is activated by a pulse while it driver is used to amplifying the computer pulse into high voltage pulses for running the motor, okay and the last type of motor we are going to talk about is DC motor well there and the last type of motor is brushless DC motor or bloc motor.

Well again DC motor is similar to the stepper motor but the basic difference between it is what a basic difference between it says that it is a brushless DC motor means that the functional effect of the motor is similar to DC motor. The mechanism is similar to a stepper motor, but it is used for very high-speed operations such as the rotors of the quadcopters all the rotors of the quadcopters are only connected by bloc motor because they are the lightest they are the most powerful which in terms of speed and they are the most simple type of motion. They are the most simple type of motor for high-speed operations they can be they are controlled with the help of pulse width modulation and specially designed drivers you can say that electronic speed controllers in the term of quadcopters they are also connected with micro mantra mostly are 32 bits for controlling their operation.

SIMPLE DC MOTOR

We are going to learn about the motor and how to control the motor with the help of rd first of all legislator content or content is a type of motors it can next content will be single DC motor, simple DC motor servo motor then after we are going to talk about stepper motor and in the last three tutorials we are going to learn about the controlling nature of the motor and we are we will be experimenting on controlling the motors. Okay, let us begin to our first topic. Before starting type of motors I want to tell you that about a motor is an electromechanical instrument that takes electrical power as an input and gives out in the form of mechanical powers it there are two types of the motor in the world the AC motor in DC motor in Arduino, we are mostly be talking about the DC motors of low voltages.

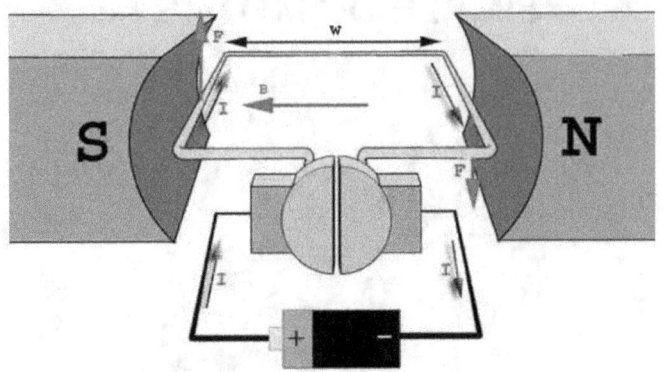

Okay let's just start about the DC motors and their types first of all the DC motor it is a simple motor would do why you can see here it has only two wires you if you want to change the direction just change the polarity one connector to positive and another negative and if you want to change the polarity just reverse the wiring and it will change direction okay thighs is a DC motor okay next what is a stepper motor? Well, first of all, it will what is a servo motor is a special type of motor that is used to view the values in degrees it is not made to rotate continuously in 360 degrees while it is meant to rotate for a specified degree. If you set the motor to run at 30 degrees it will get locked itself or 30 degree it will log itself to 30 degrees and we know do try to change the direction of the motor. If the program is running then it will set Except Saturday, thighs are the very important motor for making toy aero planes in red. Because thighs are the very important motor for making toy aeroplanes because for the rudders if you want to set the value for 30 degrees and that will be causing another effect auto suppresses down while the motor will turn itself back to 30 D so that it can move the urban very easy. The third type of model we're talking is here is a stepper motor, stepper motor Are we the third type of motor we are going to talk here is stepper motor, the stepper motor is of two types they are unipolar motor or Viber. They are specially designed motor that can run on computer commands such as auto motor, but they are used for more than 360 degree run they are useful long run especially in CNC and 3d printers. They can be very high-speed motors they can be a very low or very basic motor. They have a simple running mechanism which is said as a coil inside there is a magnet now Magnetic is rotated with the changing value of the direction of each individual coil which causes the motion in East magnetic coil is activated by a pulse while it driver is used to amplifying the computer pulse into high voltage pulses for running the motor okay and the last type of motor we are going to talk about is DC motor well there and the last type of motor is brushless DC motor or bldc motor. Well again listen motor is similar to a stepper motor but the basic difference between it is is what a basic difference between it says that it is a brushless DC motor means that the functional effect of the motor is similar to DC motor. The mechanism is similar to a stepper motor, but it is used for very high school operations such as the rotors of the quadcopters. All the rotors of the quadcopters are only connected by bldc motor because they are the lightest, they are the most powerful which in terms of speed and they are the most simple type of motion. They are the most simple type of motor for high-speed operations they can be they are controlled with the help of pulse width modulation in specially designed drivers you can say that electronic speed controllers in the term of quad copters they are also connected with micro montura mostly are 32 goods for controlling their operation.

TYPES OF MOTORS

Simple DC motor. It's working its diagram and how to connect it with Arduino to get the motion. Okay, first of all about the DC motor I want to tell you their DC motor is the most simple type of motor you have seen because thighs motor is a very small device. It can be found in toys as you have seen here the photo thighs motor has been taken outside of the toy. And here their DC motor, you can see that in the video is the B motor. They both have a different design but the thing is they both run on the same principle. When the power is sent in one direction it will rotate in clockwise and when the direction of the power is changed, it will running integrable. First of all, the DC motor cannot be connected directly to Arduino it requires a specific driver or a transistor to run. Because if we connect the DC motor directly to Arduino, it will draw a lot of power that can eventually wander and you know, sometimes the laptop also can get harmed because of thighs. So to control a DC motor, we need a driver or a transistor set up to run it. Okay, let's just see about the DC motor. Here you can see the working principle over DC motor in a DC motor you Yeah outside the body there are two magnets such as the North Pole and the South Pole okay and inside it here is the coil whenever the charge is passed across the coil or the contact center, so, the motor here you can see these are the pressures of the context and thighs is part is called the commuter is a copper ring that is divided into two parts, whenever the current is paused here it will force the motor to it will force the wire of the coil to go into one direction thighs is because of the law, thighs are because of the right-hand rule of the motor where it was driven that that was given by Fleming okay thighs is known as Fleming's right thumb rule. You can see here let me make a correction for you. Thighs is the right thumb rule. All three fingers are in other directions. Okay, let's just see it in small working. Here you can see that whenever charges pass The motor will rotate in one direction or when it comes in the middle point here. So the contactors changes direction and thighs work as a mechanical rectifier for the DC motor to continue its motion in one direction. the important part, the sleep the split ring is a cooperative that is divided into two or more parts. Thighs is the main reason for the motion of the motor in one direction because when it comes here at the 90 degree, it automatically changes highs pose in the mechanical rectifier form and thighs will give the motor a continuous motion in any specific direction as the current has been flown to the direction the direction of the motor can we change EQ here, minus start and it will be given here plus j then it will automatically change the direction from here to here.

Principle Of Operation Of a Dc Generator.

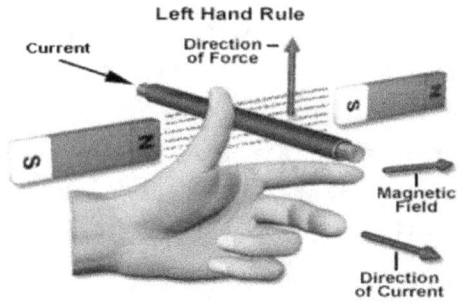

Now let's see about how to connect a motor to the Arduino here you can see that it is running on the left thumb rule yet I have shown you it is running on the left Some room here, thighs will be causing the direction of the motor

Okay, let's just see about the drivers here you can see that thighs are the driver that we'll be using here for running a motor, okay you can

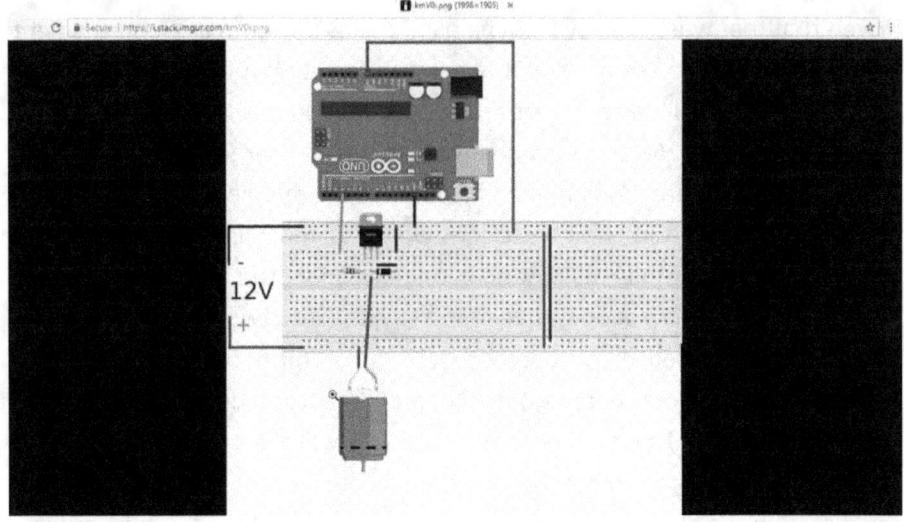

see that the diagram that is connected here is the schematic of an Arduino and a running motor with the help over 12 words outside source. Here you can see that when the transistor is connected that is an NPN transistor and about the wiring, first of all, to activate a transistor as I have told you, you should be able to give it value digital high digital right that thing number and should be given a value of five that it will make thighs true transistor to go up and it will eventually connect these two circuits giving it a closed loop and that closed loop will be taking the power source from the 12-volt battery and giving the motor a specific running output. So by thighs method we are operating the motor with a 12-volt output but the control of the motor is done by The Arduino in the interface between them is the transistor because the transistor controls the power direction or you can say that it is like a wall in power is running like water.

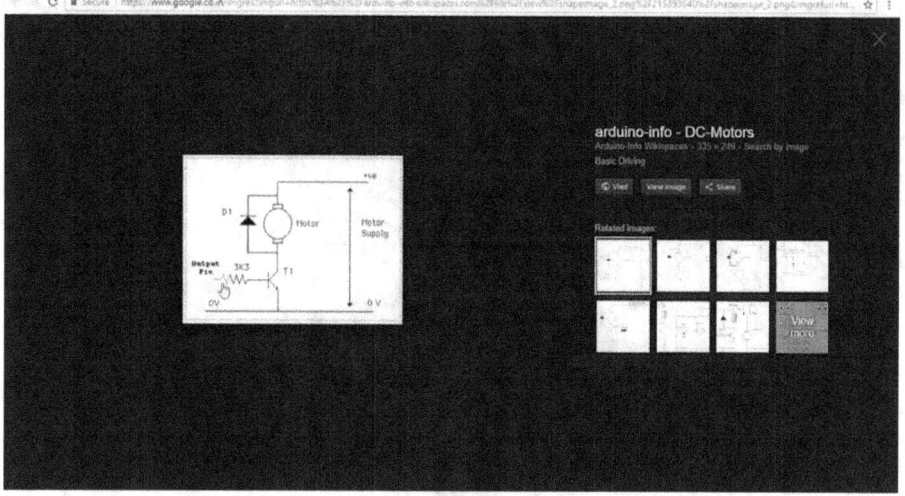

So, the transaction controls the water from the transistor controls the direction of the current or the flow of the current to give the motor rotation, but the main problem of thighs method is the motor can run in only one direction and for running the motor into the direction you have to change the polarity of the motors okay and thighs was the description of the DC motor if you want to see I will be giving you the diagrams here is the diagram and here the transistor connection and thighs side will be connected to Arduino thighs side will be connected to the power supply is required and thighs here is the motor these are the terminals of the motor. Okay and these are the basic theory of the running of the motor. Well, I think the thighs method is quite useful for running an Atlantis also but for that, we will be talking in the last section or the 10th lower running heavy motors by thighs method.

SERVO MOTOR

We are going to discuss servo motor. Well, what is a servo motor, first of all, I want to tell you that servo motor is not like a special is not like a simple motor will rotate continuously in a phased direction while on the other side servo motor is a kind of motor that is fixed for a specified value, it can rotate only from zero to 180 degree and it is not possible to rotate it more than that in many servers? While some servers can rotate complete 360 degrees in continuous rotation. But the main pros the main use, but the main thing that the servo motor is used that the servo motor can set a specified degree, if we set a servo motor to more 30 degrees it will lock itself at 30 degree in any condition that will occur to Moses or any condition that occurs to the motor servo motor in any direction the servo motor will refuse it and it will come back to the position, it will come back to the portion that is set towards example 30 degrees. If you set a servo motor to move a 10-degree direction in every second it can do that like that 123456 and up to 10. If we want to set it to five degree and exact five-degree precisely, it can get locked on the five degrees. Well, there are many applications of servo motor example the toy plane helps the reader the trader outcome control by servo motors. The reason because of that is when the servo motor muzzle it get locked into position and when the air flows over the wing, it tends to move the wing in backward direction opposing the servo while the servo motor get back to version and get the values Clear I have got some I have got some document on the internet which can help you to get a brief knowledge about servo motor okay see here our the servo motor there are many shapes and sizes Some are industrial motors that are capable of rotating continuously to 360 D well the servo motor has a basic design see

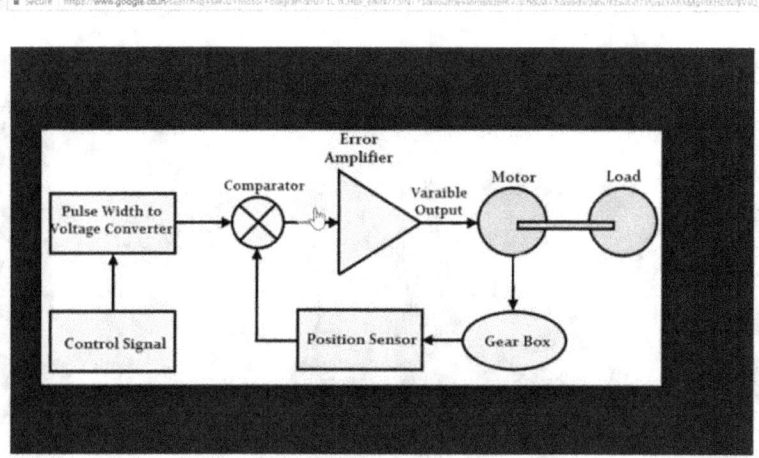

the design carefully, thighs are the control signal or you can say thighs is the Arduino, the pulse wood converter we have talked about it is a DC. Now, coming here thighs is the servo motor part thighs is a comparator a small tip that is employed that is implanted in the servo motor and thighs is the position sensor you can say a simple potentiometer Thighs is an inner amplifier a part of the chip and thighs is the motor. Thighs are the motor access when the Excel is said to rotate a 30 degree it will compare its position here and it says I will position it as zero degrees so it will give an error here and the error will be amplified so that setting the rotation of the servo motor from the Whoa look to get back to 30 degrees when the motor gets to 30 degrees the data says that there is no error it said it wanted 30 degrees and it is at 30 degrees so, there is no error. So, the value that is set here a burden or two more 30 degrees has been logged here by the access and moving 30 degrees. Whenever a change is made in the server it gets back to its location it gets to thighs point, but now, we have to discuss about the how to connect a servo motor to a Arduino. See here you can see that there are just a second there are three pins in the servo motor, they are as follows. The first one is yellow that is for signals the red one is for five volt output in the black one is for power voltage ground. Let's just see it in the diagram I have made a simplified version of it. Here you can see that the yellow wire is for the signal that is connected to the analogue pins of the Arduino.

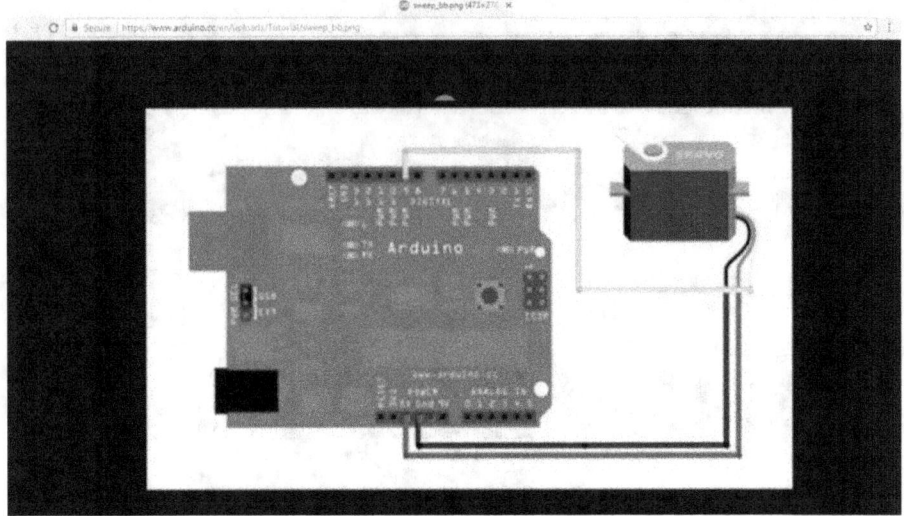

The five worlds are connected to the board and the ground is connected to the Round pain, but may I need to mention something easy if possible give the server or individual power supply because if you connect the servo motor directly to the board and the motor which has a size like thighs will create a great disturbances and it will search a large amount of power which can eventually fry the board and it can also cause the damage to the computer so, to move in continuously a servo motor of heavy torque and heavy motion, it is recommended to use external power supply and connected to the ground and then send the signal from the Arduino for the motion of the servo motors tip stability and for protection of the servo motor and the mother. You can say the Arduino board Okay, thighs was the basic description of servo motor in the following videos we will be using a servo motor and controlling it with Arduino with an example. well about servo motors, I want to tell you that there is a built-in library in Arduino that can control the servo motors. You can see here from file going to examples and going to examples and go down you can see that there are two programs built-in for servo motor application. Here you can see that the default library of a servo motor is called the library is given an object and the object is selected for all the operations. Here the object is said that the servo motor is attached at PINnine of the Arduino board, okay, the PINnine will be giving the pulse width modulation stroke signal for the servo motor to rotate. And here you can see that Look pin a PINyou can say putting an analogue pin is called a PIN. just edit. You can see that analogue pin is called here you can edit here you can see that and then over here you can see that analogue pin is called at PINthree and whenever the reading at the analogue pin has changed the reading at the servo motor will observe the reading or the analogue pin and give out a direction of motion and the angle that is that is here you can and give out a value you can see that the value for the integer is defined by the function using by the using the function map. Here you can see that in math the value of digital in the here you can see that the value of the analogue input is given as 02 1023 and the motion of the servo motor is from zero degrees to 180 degrees here when the value of the potential meter you can say that it is 255 or 555 it can take the servo motor to an angle of 90 degree if the value of the potentiometers changes the angle of the servo motor, the angle of the servo motor also changes and the angle writing is done with the help of the function. My servo which was the element and is done by the function dot write and value is we have values the variable which is the value that is done by thighs map function. So, that it can give you value at which the motion of the servo motor has to take place and at the value, at which the servo motor has to stop.

Okay, thighs were about a sarong ballistic about another lecture. Let's just talk about another program of the servo motor. Here you can see that in sweep Sweep is a much easier program because it used directly servo and later to rotate from 180 degrees to zero degrees and vice versa it does not require a potentiometer for its operation, it's simply connected to pin nine then it uses a for loop for the variable position here you can see that the variable is increased from zero to 180 degree with a one-degree increment at every 15 milliseconds Second if it reaches 180 degrees, then again when the position value increases to 180 degrees it is taken to zero with one-degree decrements. And all the values of servo motor has been written by in the object to servo motor by the right function as pose and highs degree has a difference of 15 milliseconds. Well, these are the basics of the servo motor programming and servo motor there how to control it with Arduino. Well, thighs were the structure of the servo motor that we have talked here.

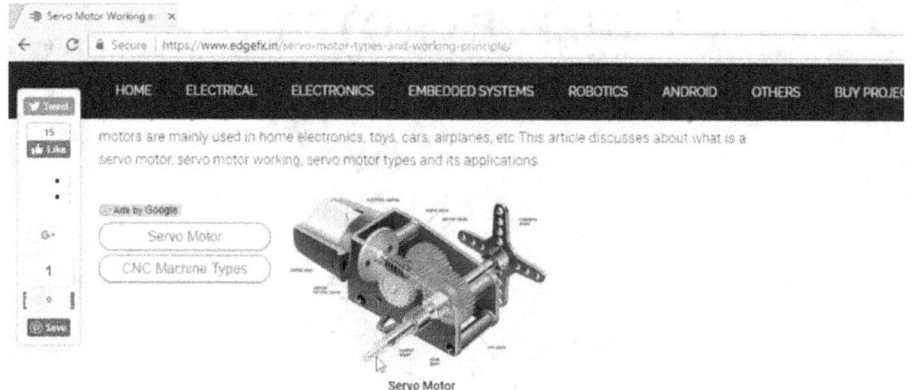

Types of Servo Motor

It is using a more simple motor and there are some electronics behind it with a potentiometer and these are the gear timers hear a potentiometer is connected which gives the value at which the servo how to stop or at which the servo how to reach there is done by the value employee the value correction has been done, the value correction is being done by the comparator which is pre built in the servo motor and the value for the signal input is taken with the help of Arduino So, that the Arduino can command the servo motor to get to a specific degree. These are the different types of servo motor and here you can see that an AC servo motor which is an industrial servo motor, which has a petition positional encoder rather than a potentiometer to give very precise values so that it can perform very accurately here it is a similar servo motor that we have taught it is called continuous servo motor because it can rotate to continuously 360-degree Indian rotor to many times you can see that here. You can see here that it can rotate continuously. These are the spatial servo motors that do not have the pin that stops They do not have the physical part that stops them from rotating continuously and they are potentiometer are of good quality so that they do not deteriorate the potentiometer will run into continuously 360 degrees.

STEPPER MOTOR

We are going to talk about stepper motors. Well, a stepper motor is a special purpose computer motor that is used to run at specific degrees as well as for 360-degree motion. Thighs are the most common motor which is used in CNC and is used in 3d printers in two dimensional protests. These are generally heavy industrial motors that can run Every load as well as they can rotate in specified you can set to rotations or say certain 20 degrees it can also move to micro-stepping So, that it can move to one 10th or one by 200 in many spatial motors of a degree, you can say that a stepper motor generally can rotate only about 200 steps out of 360 which makes one step about 1.8 degrees, but if you use the micro stripping method to control the servo motor, it can move about one by 200 the part of a degree which can give a very bright clarification of the servo motor and can give very accurate results. Okay about the servo stepper motor about about the stepper motor, the stepper motor comes out it the stepper motor comes into aliens. They are the unipolar motor such as thighs and they're bipolar motor such as thighs. There is only the difference of wiring in you can say they're a bipolar motor like thighs has Oh four wires while a unipolar motor has five wires. These are the difference in polarities of the coils that are inserted in the servo motor for direction motion, direction and motion. And there is also a hybrid motor that has six wire they can run both as a unipolar as well as bipolar. Okay, let's just see a diagram so that we can get a clearer understanding of what is a stepper motor. First of all, let's get it from starting here you can see that the red part is in a magnet in the 123 and the four there are numbers they are the cause that activates thighs that activates by getting the value from the computer and get amplified by the driver.

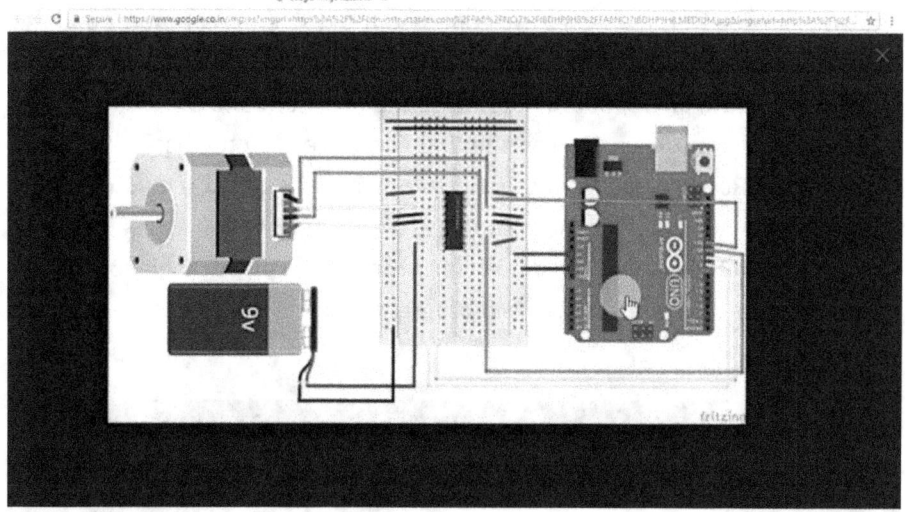

So there they can create a motion to the dry glow magnetic drive or you can see that the Excel of the stepper motor is made out of magnet for giving it motion okay thighs is the method for controlling a stepper motor. You can see that here in the diagram whenever the pulsating clocks in it evolutionary will or the array is called in incrementing direction from 1234 it gives a small value of motion to the rotor and it goes continuously like thighs in motion it can give continuous motion, but very precise value, if you use magnets dropping the value, can be precise even more okay servo motor and bouquet stepper motor There is also a special type of stepper motor known as bldc as we have discussed earlier there is the same principle of working in dc motor The main difference is the outer coil is replaced by permanent magnets while the inner core is made up of pulsating magnetic coils and they are controlled by high-speed AC or you can say electronic speed controller by microcontrollers and they are all 32 bit Tokyo high performance and high speed of values so, that they are used in quadcopters okay about the wiring. Let let me give you an extra Sample let me give you an example of a stepper motor both as unipolar like thighs and bipolar like thighs. You can see that the thighs motor is bipolar because it has four wirings that all go through the driver here they are using a simple chip for the driver and the driver or you can just signal amplification, the amplification power is eating by a nine-volt battery in the signal we're running a motor is gating by Arduino okay. While thighs function is very similar to that is used in unipolar motor, it has a driver and you can see that five wires are coming out of the unipolar motor and the power source. Here the power source is used by the Arduino but for better performance, the power source should be independent and protecting the motor from trying. Okay, you can see that there are they the basic difference between the Romanenko motor, bipolar and unipolar is that the vote drum on different types of codes there coding is different and they have a very similar motion they can be used both for micro-stepping but their source code and their way of algorithm and programming

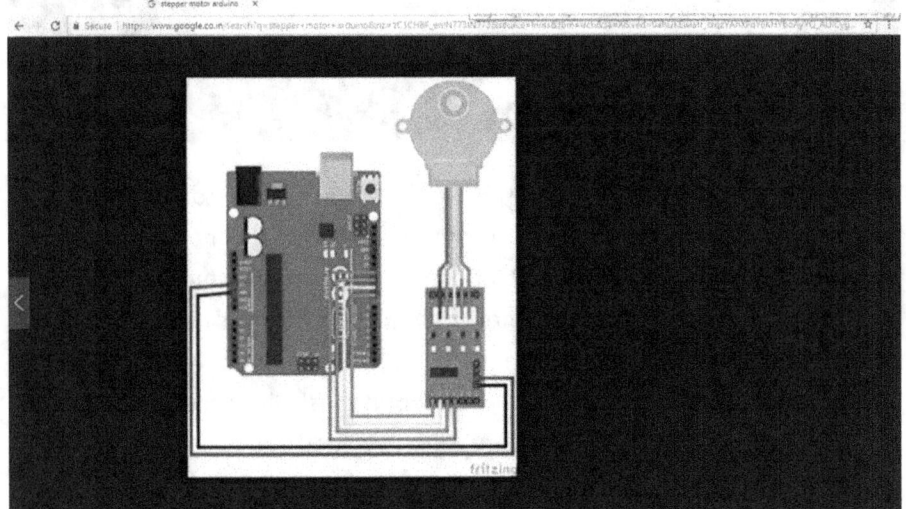

is very different from each other. They mostly we are talking about the bipolar motor, mostly we are talking about the bipolar motors become bipolar motor is the most common type of motor we are getting in interfacing in

creating an interface and to get an operation in the further future chapter we will be talking about wiring a stepper motor to Arduino and using with a driver to aid in the signal amplified in learning the servo running and running the stepper motor in the future chapters. Okay, thighs the stepper motor library is already pre-installed in Arduino.

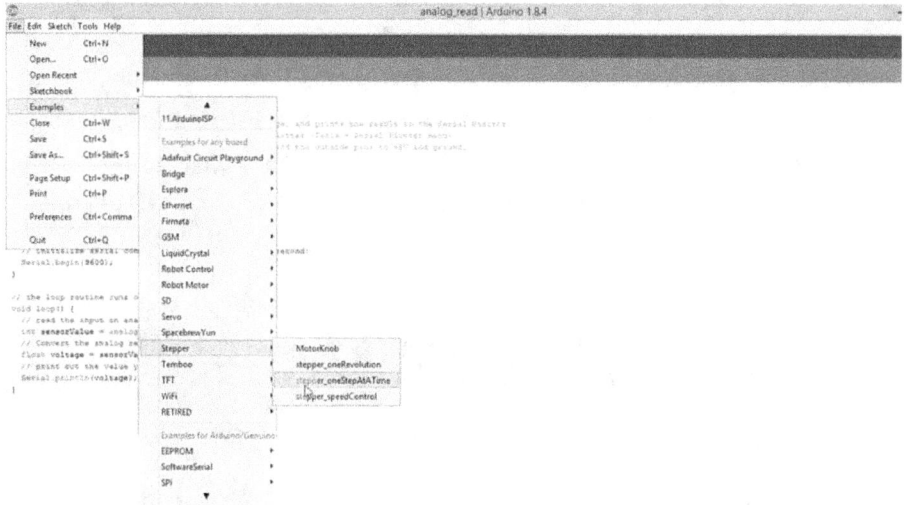

The Arduino has built-in libraries for the motion of servo Motor as well as for stepper ledger see them here going in when you file example you can see when you go down you can see the library design for stepper Well here you can see that motor norm one revolution one step at a time in speed controller just see them.

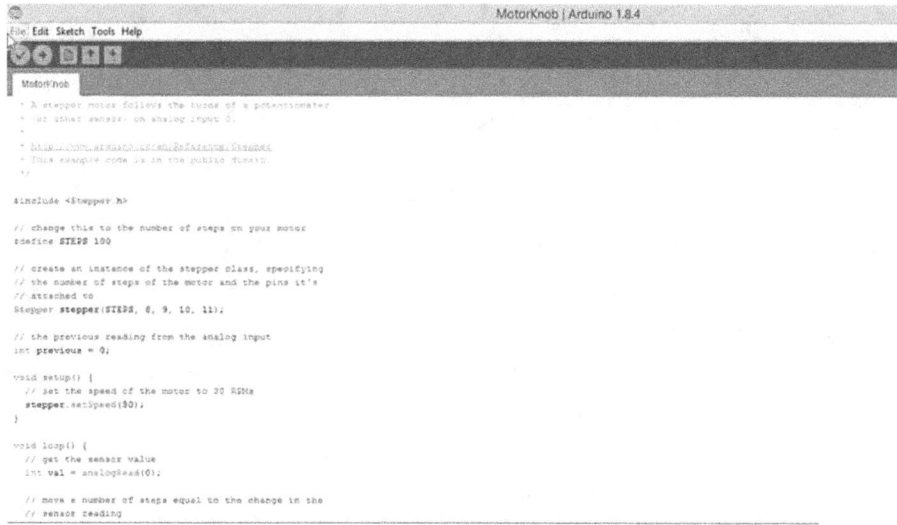

First of all motor home motor node is a library that is used for running the motor. Or you can say that it just defines the PINin which the Arduino can be connected, let's say eight 910 11 and the variable there is the steps. The steps are the variable which is slow the number of steps that have to be taken by the servo motor to reach a specified value. Here it is taking 200 steps, it means that the motor will rotate from zero to 180 degrees, you can manually change the steps as to P 60. Or whatever you wish. Now let's see another program go-to example. And again go to Stanford you can tell one revolution.

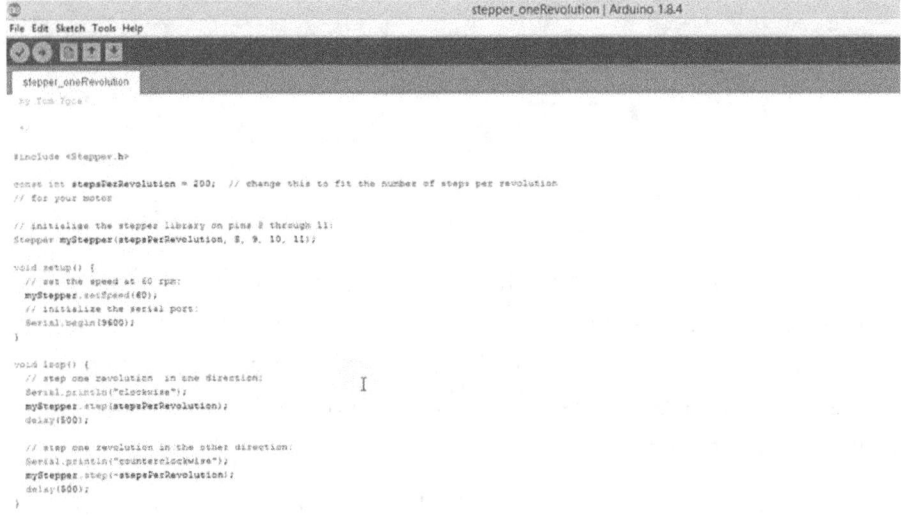

One revolution is thighs small program we just say that the PINs Motor the motor which is connected to the PINeight 910 11 we have to route it all, all the 200 steps to get back to get a complete revolution motion and come back to its original position where it has been started, the motion can be tamed or can be controlled by controlling the speed in thighs part, we can see that the motion is having each step time difference has 500 milliseconds or half seconds, you can also write them as 50 milliseconds so the motor will run about 10 times faster than it but do you have to change the time direction of the board, you can you have to change the time value of board the variable so that it can give a clear rotation. If we change the value for one, it means that the motor will have wounded evolution in one second and outer gating that it will rotate 10 times faster to graduate to initial version so it is better to change the value of both places so that you can have smooth motion in both directions Another legacy than other part is motor one step at a time thighs is a program that is mostly used for taking the step and is also used for getting the precise value of the stepper motor.

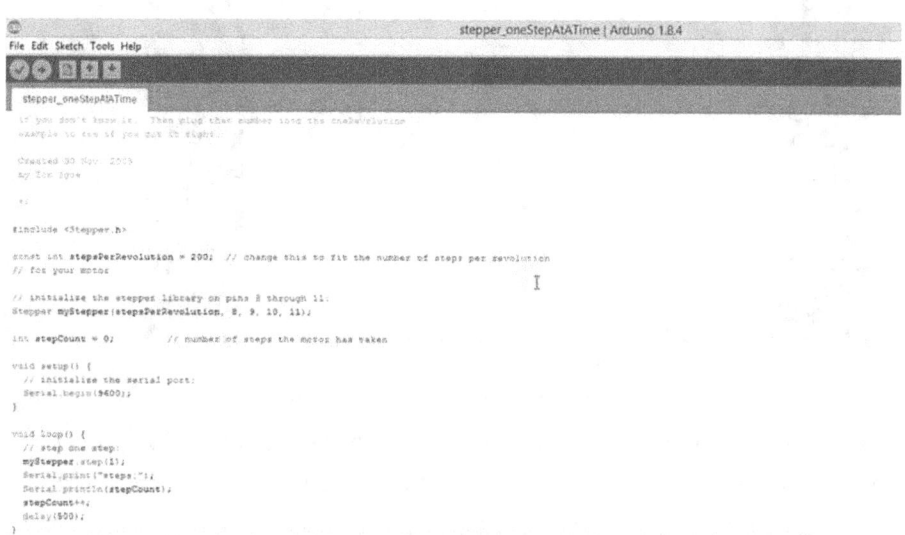

Thighs program is used for getting the stepper precisely useless that you can get the interval of wider interval you can get that by the interval of one step you can see that absolutely how much angle that the stepper motor has built from the previous to the present location. Actually, it is 1.8 degree but sometimes it can also change because of micro-stepping you can change 1.8 degrees to 200 times clearer or you can say it you can divide it into 200 parts to get a better-clarified value. Well, thighs method is used for controlling of the stepper motors that are used in 3d printers or CNC protests, because they require a high level of accuracy and for higher accuracy, for a high end for the high level of accuracy. The stub degree should be very precise. So they're using micro stepping, we will be talking about micro-stepping afterwards, but let me give you a small glimpse about that stepping can be done in three parts. Let me show you the difference, Okay, you can say

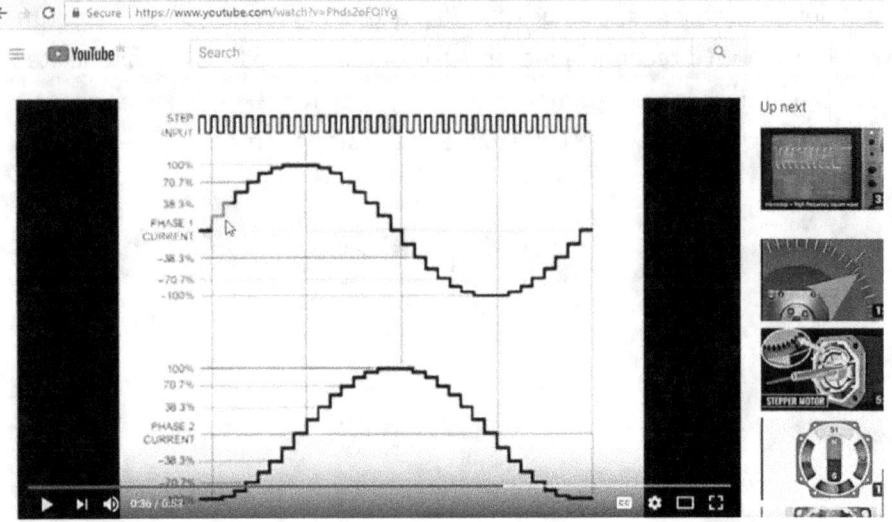

that it is same as pulse width modulation if the pulse is wrong, then it can do a degree but if the pulse is very slow or you can say that the duty cycle is very small, then it will give the value to motor node one degree one point node it will not give the motor to move 1.8 degrees but it can also control so, motor two half a 1.8 degrees to one-fourth of 1.8 degrees anyone the 200 the part of one two and eight degrees for that process the microstamping is used. I will be showing a diagram with for a second here you can see that the example of micro-stepping Thighs is a complete duty cycle that has been divided into four basic steps for you can say that each has a difference or 25% of duty cycle, but thighs steps can further be broken into small parts here you can see that thighs are the method here you can see that each part is broken into our seven segments or the So, that it can give a value of one of four into seven that will take you over 28 one by 28 the value of the stepping that should be done so, it can give a very precise step to be taken. The whole cycle is for 1.8 degrees present it can give 1.8 by 28%. At present, it can give 1.8 divided by 28 and that is the angle. The micro-stepping is the most precise step that that is The most precise step can be taken by the help of a specified driver. But remember one thing for micro-stepping it is almost, but remember one thing for micro-stepping it is always use a 32-bit board as well as a 32-bit stepper driver to giving verifiable use in industrial strength in, in industrial hardware, all the stepper motors are being controlled with 32 bit ARM architecture microcontroller and 32-bit stepper driver. So, that they can give very fine precise value and they can run the operation at very high speeds. Okay, thighs were the theory about the stepper motor here you can see the Arduino the wiring thighs driver, the external source of battery and the stepper motor here. We have talked about the program in the chapter we will be running a stepper motor for practical purposes.

RUNNING A DC MOTOR

We are going to talk about running a DC motor with the help of an Arduino board and driver, we cannot run a DC motor directly with the Arduino because DC motor draws a large amount of current in thighs current can easily fry the Arduino board even in your computer or laptop.

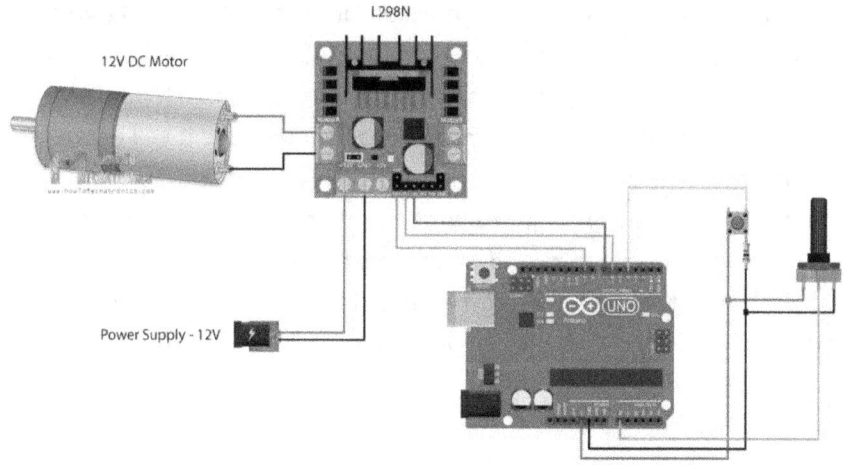

So be careful while connecting the DC motor to the Arduino because there should be no direct connection or there should be no leakage voltage in the world. So there should be no leakage voltage or current in the water which can directly go to the computer or the Arduino because it can burn it severely and cause you a lot of damage. Okay, so protecting the Arduino from the motor in the car having good control of the motor we need to connect it to a driver. A driver is specifically designed a machine which is used to perform the function of controlling a motor via the help of small signals small digital signal data provided by the Arduino and by amplifying the signal with a built in voltage sources here we are connected to here we are connected connecting it here we are connecting it to a 12-volt source so that the driver can amplify the signal in for the control signal and for the control signal we are connecting it with Arduino and the output of amplified signal and control signal is going to the motor for that we can select the direction of the motion as far as we can perform the speed variation of the motor I have selected the analogue pin for motion control and insert into a low so that we can have a clear vision of the motor running Okay, let me tell you about the diagram. Here it is the Arduino it is coming out with three in three themes it is coming out in the form of three things. The first thing is the analogue ping which is used to control the speed of The second and third pin other deleted pins which are connected to detail pin four and three in Arduino and are connected to the next loop and are connected to the next two pins on the driver. The first two brackets are connected to a 12 volt in the motor is connected to the output terminal, we can change the direction of the motor as required by changing the PINvalue from high to low. We cannot keep both the pin high or whether we look because it will create zero potential difference between them causing the motor to not rotate in any other direction. For moving a motor we can set one value high and another low to move in one direction. And we can change it as one more value low and another high to move it in another direction. Okay, let's just do a practical

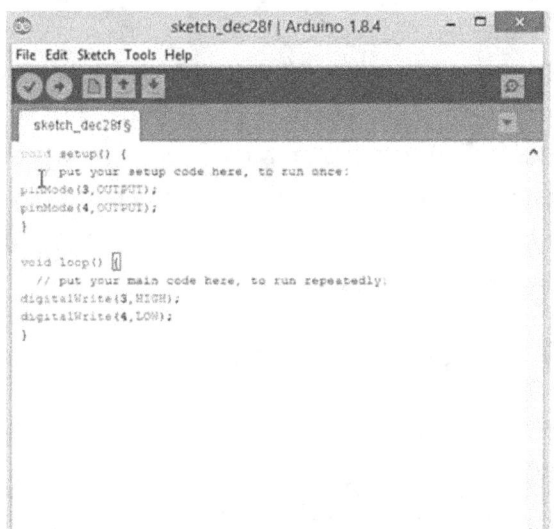

open Arduino opens a new sketch and right here. First of all, we have to declare the pin right we model let's say PINthree comma output. We have done PINthree is out Now again pin modem od e emote cepi number four as output because I have connected it to that pain if you have connected it to another bin please please be careful by writing the PINs Okay. Now come to word low write a simple program say digital write in number three comma H I g hand we have to say p number four as low okay right in next line big it I w r I t detail right p number four Komal or w, Okay, just take the program in more three and four I declare as our port The port is connected to PIN10 for details Right peanut three high painful low and the motor is here wiring is going correctly in the please check the wiring because it's going to severely damage the microcontroller if wrong wiring is done at high voltages okay let's just compile it we are compiling it okay just cancel it come here it compiles it's compiling Just wait a second okay there are no compilation errors go to upload and upload the world values, okay the world is uploaded and you can see the motor is rotating in one direction. If you want to change the direction of the motion please go here. Say the PINthree as well Oh, we are experimenting both in number three and four our status low not check it can the motor rotate on both low you can see there the motor has stopped working now stay thighs law and for number 10 as high g h o Again upload it to the world. Now you can see that the motor is towed now you can see that the motor is rotating in another direction. Well, thighs are the motor for thighs, thighs are the method for controlling the direction of the motor, but if you want to see it in on the screen let's just write a simple program. First of all, I have to stop the motor for that it will not create any disturbance in our programming I had to stop the motor. Now if you want to see the direction of the motor, it is good for us to get a clear view by using it in getting the direction and serial monitor right is E RI I dot veg I m serial dot begin right 9600 for the baud rate, we have given it 9600 for that function. a mega simple statement you can say that thighs are an animal form of if statement. Okay, let's just do it. It is taking a second. Wait, stop.

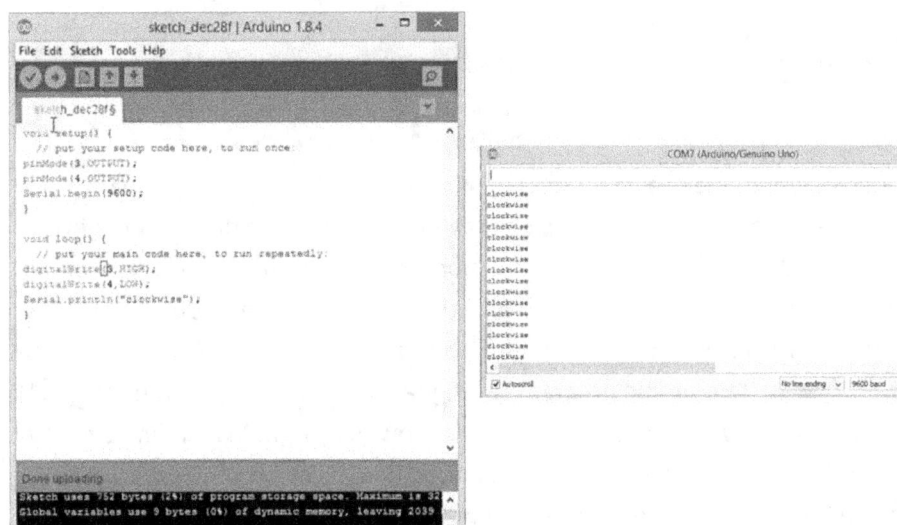

Okay, come back to programming right here. Clockwise serial SPI I dot p r p I n d Im for the next line. Okay. Right is blockless Okay, we are done that said one thing too. High, I'm sitting with number three. Now upload it to the world. Cancel it, I am cancelling it. You can save it as you want. Now see the direction of the motor? I think that's clockwise. It's going clockwise. But what if we want to rotate it for 10 seconds clockwise and 10 second anticlockwise? Well, there is a method for it. First of all, let me stop the motor. I will show you the method I'm stopping the motor is taking some time okay and it stops there is a very simple method to rotate the motor in both directions just come here if we want to rotate the motor in one direction for 10 seconds right the program d e l a wide delay let's attempt again. Okay, stop it. Now come here right di JT LWR at Digital writer in number three, SL or as l or w comma digital, right? Lu our ID VA big idea my mistake retail right? For coma, it gives High,

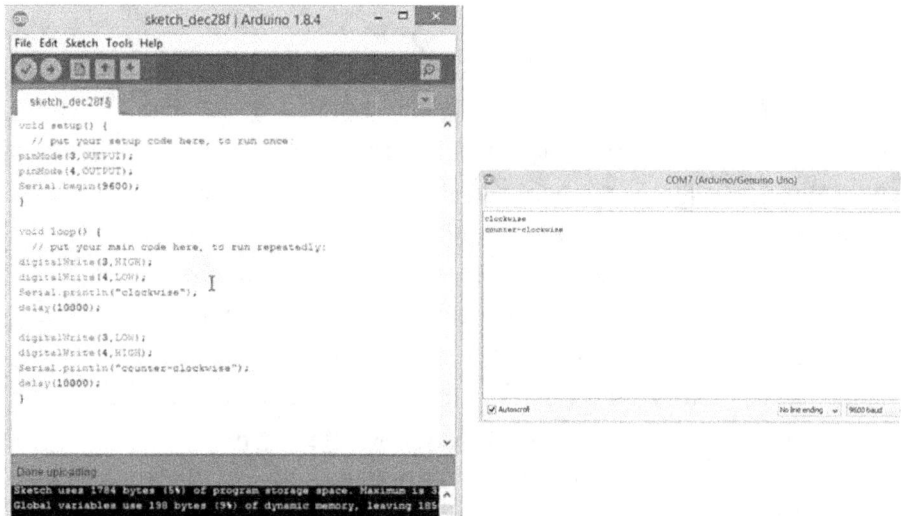

come here, you can see that thighs are high for that moment for 10 seconds and it will be rotating in another direction for another 10 seconds is our daughter PTL n print In you can say it counterclockwise or counterclockwise terminate the line and write d a y delay of 10 second little 10 seconds 12344 values Okay. Oh now give the I'm giving them water for your understanding of loading the program. And let's say with DC motor running. Okay, so in the program, you can download all the programs on the starting of the module in there you can get the program Okay, it's taking some time. Okay, let's just open the serial monitor. Let me open it once again. Now you can see clockwise for next 10 second. It's running continuously, just wait for a second to come. And you can see that it's rotating in another direction. Well by thighs method you can control the motor in both direction and you can get the value of clockwise and counterclockwise by changing the polarity.

RUNNING A SERVO MOTOR

Well, a servo motor is a special type of electrical motor that is used to give a precise degree of value. It is not used to give a continuous rotation well to connect a servo motor to an Arduino It is very simple just you can do it by connecting the yellow wire to signal the red wire to five volts. You can connect it with anything yellow wire to signal or you can see the pin which we have defined for controlling the servo five-volt wire will be connected to the red wire will connect two to five-volt output in the ground wire should be connected to the ground terminal of the Arduino. Yes, it is good for running small motors but if we are using weak servos then it is necessary to view them individual power supply. Okay, now let's begin it with the completing a core. Okay, come here. Yeah is a code that is a library is always all a good library for controlling servo motor is available in the Arduino before but we want to write our code? So, we can do that first of all go here at the top include we are calling the library I we are calling the library by writing include I nc LUD included will change the colour it means it goes servo rotor resistant It included you s e r v o servo dot h s is the extension for c file Okay. Now we are rights e r v also a row is going to get an M m y m So, or you can just say my that is good. We are giving high my Okay. Now, in Word setup we have to define The positions that should be in mass or vo go to the top we need to declare an integer for the value storing of the angle that should be taken by the step we are declaring I nt in GL in GL is not possible so let's just say D eg entity eg it means degree okay is equal to zero. Now come here said m while right Empire daughter a TT SC, h is we are defining server p number three. So we are calling in Why don't attach the number three service-connected PIN. Now, it is very simple to control the motor for 32 at a fixed location or 30 degrees, simply write my Dr W r I t my adult right to say 30 I will be using are variable but right now I'm just trying to be as simple as possible. Okay, the last program is running. Just hold on a second but the symbols will be removed from here. We are in declaring. Now uploaded. Here we have to define my forward setter, Okay, come here, cut it right here. Okay, let's just upload it to demo monistic. Let's return it. It's compiling for a second. Just wait for a second. It will do it. And it's the servo has been taken the position of the 30-degree legislature to for TD

okay, it changes the value now legislate to 19 ads not it's a Kindle edition it just says yo degree okay let's just say quantity D that is the last value that can be defined using the servo. Okay, but what would happen if we try to move it is not possible to move it because it is locked at the position Okay What if we want to see the motion of servo very slowly you can just go to files open example go to servo motor see nowhere select the penis PINthree, four minutes be number three you can select any envelope in which you are connected to Arduino. First of all, let me explain the code, we have called the liability then we have defined objects servo as my servo. Next, we have defined a position for it and it is initialized with zero. Next master was not attached it is declaring the pin that is taken a signal that is PINthree for my case and in Word of the loop. We have written a program folder for the position is equal to zero less than 180-degree version should increment by one step in every 50 milliseconds. In the step that is implemented should be written as the version for the servo motor in the same happens when the person reaches 180 degrees and it is greater than zero, then it will remain with every 10 milliseconds and the version will be written in the servo motor as the angle. Let us upload it to the world it's compiling with for a second. Now you can see the motion of the servo motor.

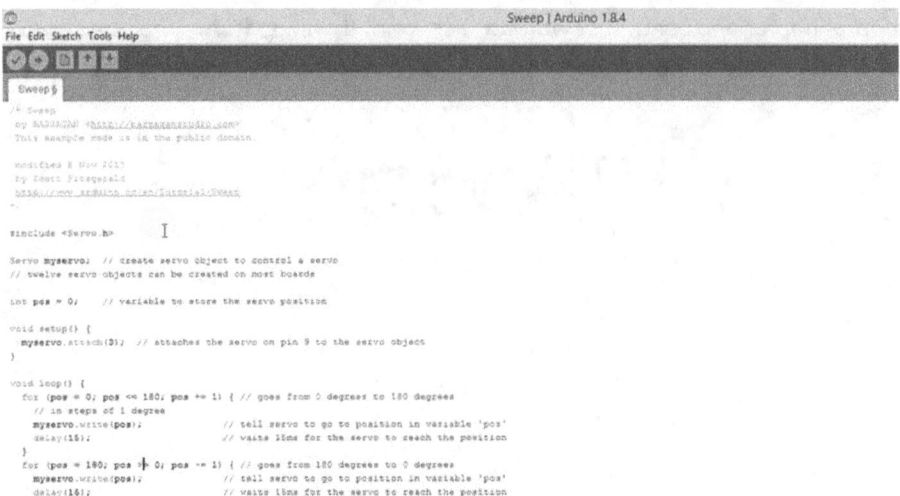

Its happening thighs is the motor of the thighs is the method of controlling a servo motor you can write it manually you can call it a program, give it a logical function or if you want, you can just test it you can just take the servo motor by thighs motor. Well, thighs were the video tutorial on controlling the servo motors. Keep in mind for running heavy servos It is better to give them individual power supply connecting the power supply to the ground also microcontroller board, then using them if you connected if you connect a V servo directly to the microcontroller it can damage the microcontroller as well as the components.

ARDUINO BLUETOOTH

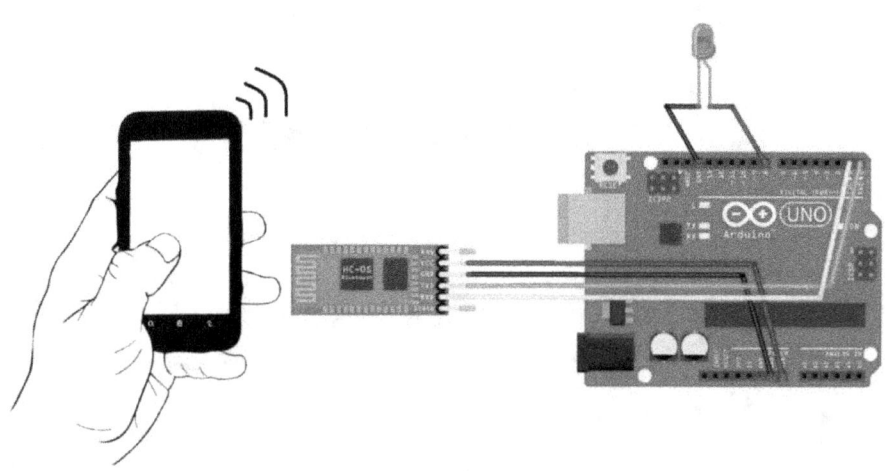

There is only a few connections to be made. So let's get started. As you can see in this Arduino board we have these sockets plastic sockets with numbers beside each of the socket. OK. Now these plastic sockets have Dunbar's these numbers of rulers and we know bins that Arduino microcontroller used to interface with the outer wall and these pins we will connect to Bluetooth much y'all and resistor and lead for testing for other bands. Benzedrine been one dot x and t expends these two ones as you can see all the XTi x 0 and 1. They are angry and one of those will go to the x or x and the Bluetooth will do all the 5 volt and unbelievable Arduino Benn's will provide Vcc or power for that Bluetooth motherboard which is here. The undie which is ground will provide the ground to the Bluetooth module as you can see here on the this black cloud will get the ground from that that we know and we'll get this sorted out when you get 5 volts to about that what you will from Arduino. So these four wires are the connection to the Bluetooth what you want. So we kind of get it up and running. Then we connect alid negative to the ground here as you can see why the positive to bend 15 with resistance value between two on to one kid. So that this let one burn out if we connected directly 5 volts will go through this lead. Why it only operates at 1.5 to 2 and 5 volts. So we need to either store to minimize the Volt the bore the sleds. Then we are good to go. You are done with this circuit but please not don't connect our X thought x and x to the X as you can see. You must connect our X we know to t x. Since R X means that it's evolved out of the Arduin or of a C or A C of the signal from that transmitter of a little with these ton's for transmitter and vice versa the transmitter and the Arduino will send data to the receiver of the truth. So each one will be connected to the other one not the very same that all the other watches are here to t and t to off so don't connect or x or x anti x 3 x or with two or three no you will receive no data here the X means transmit and R X means receive.

HOW DOES IT WORK

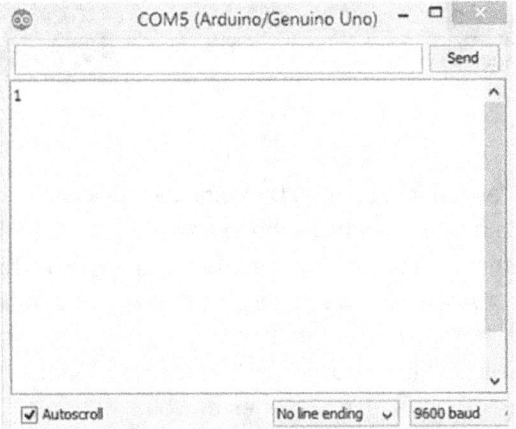

Actually 0 5 or 0 6. Depends on what you will you will have for the Bluetooth model works on serial communication. Here's the under the Android app. It's designed sending serial data to the Bluetooth module when certain button is pressed. The Bluetooth will do all other in the C of the data and send the Arduino throw that x been over with what you will which is basically are ex-Penn of Arduino the code fit to Arduino chickens that received data from that Bluetooth y'all and compares it received data is one that it turns on or if it is 0 then that it will turn off the binding on the received data in order to test this.

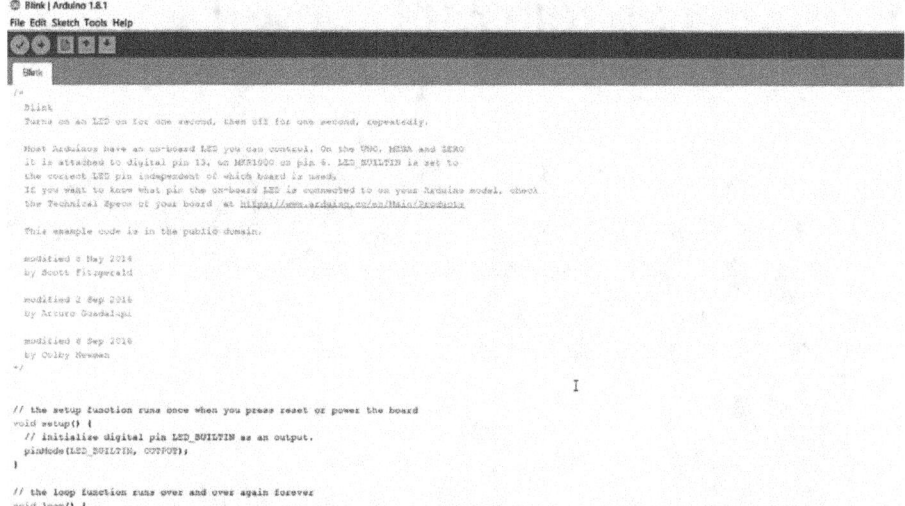

You will have to open the serial monitor and watch the received data. Let me show you the serial monitor before we start doing the experiment. Go to Start Menu then search for ID. Now as you can see here this is our idea. This is Bill from the company. You must go here tools and go to the serial monitor here. As you can see this is it. This is the city of Toronto which will act like our mobile phone and the simulation. You can also go to it by clicking on CTRL SHIFT. Plus I'm on the keyboard and it will be Bob. This is someone once we connect the green all we have to go here and choose the ball to which we are connecting the green on board. That's it for how it works. As you can see it's really simple.

ANDROID APPLICATION

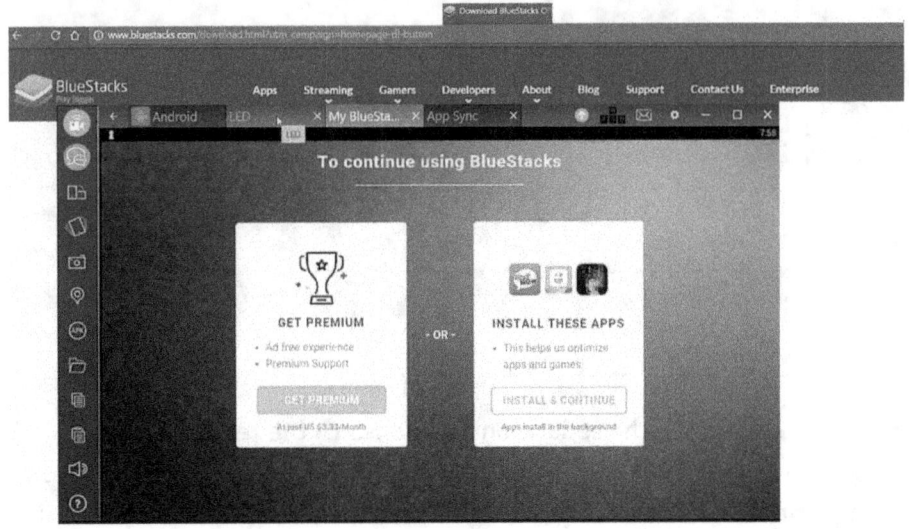

You can download the Android application from here and the source code of the entire project so that you can edit yourself if you have previous Java experience. Now how to use them don't know the application of from here from this link it you and WD tooth comb slash and slash Arduino Bluetooth which is our official Web site or from the other link. If this thing didn't work now after going to this Web site you will find the download link in this article which will be as you can see here. This is the download link you will get it from the article and you will be directed here to this very folder which is this cost folder. Now as you can see we got an ad megabyte file which is and that zip format so you need to unlock it and start downloading that farm. Now once the download is finished close this and go to where you downloaded that file right click then extract. So Arduino Bluetooth as you can see. This is our folder. Once you enter that folder you can see that we have your files. This is the Arduino code. This is the beginning and this is the zip file for the source code for this big file. This file is basically what you will download and install on your phone. Now for simulation purposes or will use a simulator of Android simulator for Windows environment. You can get it by going here bright blue stocks as you can see blue stacks is an American technology company that uses the blue stuck player which is used to simulate Android system on your Windows operating system Android emulator for PC and Mac. If you went there you can simply click here download Bluster and just wait.

Then you are good to go. As you can see here it's downloading. But since I downloaded that and install it I'll find it here. Blue stacks. That's it. Now as you can see it will take some time to start this up. This is it. This is our simulation. OB stocks it will take some time to download all. Didn't think on your computer specifications that graphic card and the Sibiu. Now once you have it once you have your desktop you can simply check these options as you can see. This one is one that we need. We need to install it. OK. That's asking us for the location. So we need to copy that location on best to tear this is it. This is our biggest find simply double click on that far and now it's installing a key OK you can see that has been installed. Just click on that to start our application which is lids. It's asking us to do something if we want to keep using Bluestar. OK. And that's thought to continue. Now going back to THE LEAD application. OK. Seems like we must get these apps installed before being able to use this software. OK. Now it's asking us and I wanted to turn on Bluetooth. Just click on allow turning Bluetooth on. This is our lead up but since this is a simulation environment we might face some problems since there is no Bluetooth. What you will for this environment. I just wanted to show you the use of interface. So let's get back after running the software and once turn and go and up with truth, you must build your device with a C. 0 5 0 0 5 0 2 0 6 2 what you will turn on I see 0 5 0 0 0 6 Luthern would then scan for available devices using a mobile phone bill to see 0 5 0 6 by entering a different password or says one two three four or four zeroes. That's it for the first step for running the software and building your Android device with the Bluetooth what you own. As you can see this is the user interface that you must click and or Bless their devices then a screen on your must to lift your Bluetooth what you will from the list which is Seasonale 5 or 6 0 6 after connecting successfully by entering the pin code 4 0 or 1 2 3 4 press on the button and then on that lid's which is connected to up we know on this off to turn off that light which is connected to either do you know when you are done. Just click on the disconnect button and you ought to go then turn off your router what your this is. There is an interface for the application. After downloading the file and installing it on your phone. That's it for this lesson about the Android application. If we went back here as you can see it's facing a problem telling the truth Bluetooth your sins. There is none here. In the simulation mode but you're simply transferring it back a file to your mobile phone and install it in your phone. Since that is Bluetooth you can simply discover that she's 051 you all and everything will be able to go. That's it for this lesson. Thanks for watching. If you have any questions please do ask and you are on the Ibar. That's it for now. This is educational engineering to.

SCHEMATIC

This is for us weigh in on all this is the USP programming cable that you will connect to your computer188 and it's also used to watch what we know in some cases if you don't have a lot more requirements this188 is the decision that you connect Bhullar 5 volt small supply for that if you need more board requirements.188 And as you can see these are the sockets that interface the Arduino on our microcontroller with the188 other for this dx and our expense zero and one are connected to the x x as you can see the X here connected188 to the X in the wall DXi is going to start in the out of the window.188 And these two wires that ground and VCC are connected like ground and VCC for the bolts for the OP on188 them.188 So Arduino is boring the Bluetooth module using these two red and black wires.188 This is a quick conversion for a schematic.188 And as you can see Number 13 is connected to this store and to the positive terminal of that lead why188 that negative 10 minute is connected to the ground of this of being on aboard.188 Next, we will talk about why we did connect this been about 13 since we would be using it in the code188 section but that's it for the schematic.188 This is the final version before going to the code.

CODE

```
char data=0;
void setup() {
  // put your setup code here, to run once:
Serial.begin(9600);
pinMode(13,OUTPUT);
}

void loop() {
  // put your main code here, to run repeatedly:
if(Serial.available() > 0)
{
data = Serial.read();
Serial.print(data);
Serial.print("\n");
if(data == '1')
digitalWrite(13,HIGH);
else if(data == '0')
digitalWrite(13,LOW);
}
}
```

So try the code we need to do you know Id just go to the man then hit we agree on no ID. This is it. And we know. First things first we need to create a new project as you can see here. This is on your project window that I and Cree are the size of this window. And let's see what we need from this Manaos OK. It's got a bit of launch. OK. We need to increase the font size to 20 so that you can OK this is good. This is the set up you can put your set up code here to run once. So between these two brackets, we will but our setup code for bends. This is the log from it then it will that be it and you need to put your main card here put on the beat tiddley order indefinitely. First things first if Pfui needs to identify any variable we need to define here this area we need one variable named data and give it a value of zero. This is the initial value. You don't need to put it so that you have control of all of the first value of that. This first of all have if you're literate without equal equaling to zero and you will be forced to get the value which is available then more. So no we need to take control of this variable so we give it a value that we know which is zero. So that change will happen from this base which is zero. Not in this area you need to click with your mouse button here and let's start the senior communication module. You only to write cereal dot begins. Then as you can see right here the baud rate which is measured in bit per second how many bits per second will be transmitted or received. It must be the same for both modules and we are using 9600 baud rate for the serial data plans mission here. And the set up then we need to assign bin number 13 to be out. But since we are connecting it to it it will take an output signal to turn on or off. Had right been more than the men on board which is 30 and then output that. We initialize the communication 9600 baud rate then we assign bin number 13 to be as output. Now in the void loop, we need to write an IF statement that checks if there is real communication is happening or not. And so that we only send data when we receive data. So but if OK this is the main form where we need to either condition cereal available more than zero and inside it, inside these two buckets we would like our code. OK. OK, now I would write of course he is in the middle between these two taxis. First, we need to read the incoming data and store the incoming value and the data about your bank data equals cereal that reads. Now we need to bring the incoming data and value inside data in Syria them on the top so that we can Chicot's to make sure that we will see the light data light cereal. Dr Brant by the variable data he is after that we need to move one line below. So if the then thought that I won't come in one line just like see it again. Brant Okay semi-colon inside that we need to lights slash on which is an abbreviation for newline and means in the new line. OK after I think of the data we want to check that this data equals one OK. That is it does it equal one. If it does equal one we need to write as we mentioned in the application where we need to turn on the lead which means that I think the value of one of the high logic in about 15 now is if data which is the value received equals zero then we need to write logic low or zero to bring in about 15 to 10 of that lead. OK, this is it. This is our code. Now we need to save it Game 7 slide or fold or it's making you fold out of your chords. Ok called. Just click on this button to verify that our code is doesn't have any syntax at all. As you can see combining Skitch OK everything is good to go. As you can see that isn't all it is here done combining our code only contains 19 lines. OK. And now I thought if you want to do that all on we just need to click here. After making sure that we choose the right to come from here. After that, we don't know we need to go to that device manager. And check the clipboard to which as you can see here which are on I was connected then we need to move the fight here. After that, you can simply be connected by clicking on the block button here. But since he doesn't have an Arduino it won't do want war. And the next lesson I would connect to an Arduin on are and we will test this code in the city on the monitor board. But before doing that let's review what we did here. At first, we defined a volleyball for storing received data. Then we set the board for the transmission. And this line. Then we said digital PIN 13 as output. After that, we send data only when we receive data. So you need to check if there is any that are received. If there is then read the incoming data and store it into that variable blend the value inside data and see the moment monitor then go on your line for think on any new data check whether a value of data is equal to 1. Now if the value is 1 then let alone if the value is zero then laid off. This is our code. It's really simple. You can change this with whatever you want but it's all following the very same principle.

TESTING CODE

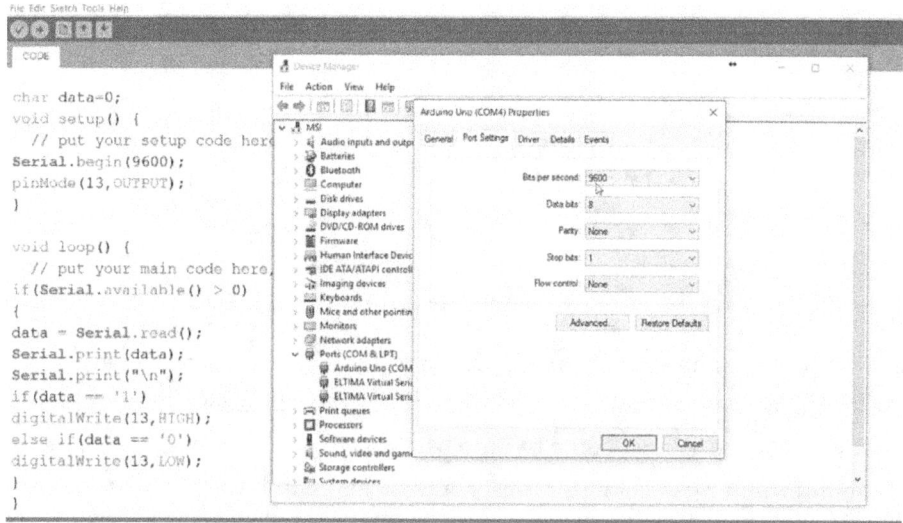

Arduino board to you or you will be bored and your computer or laptop OK for connecting it. Let's go to the device manager. To see the combo that will be assigned to us we know. So as you can see here there is no on board assigned. Now once we can nick WEO board as you can see Arduino on the on come forth. If you double click here as you can see it's board up. We see these other ball to things. 9600 baud rate. Now we need to go here and click on Chrome for all we know on and then simply go on click verify over the file called again then click on the upload button to upload it to. Ok done uploading. Now the code is uploaded to our Arduino on a ball in order to test it. Let's go to Tools Siri on monitor.

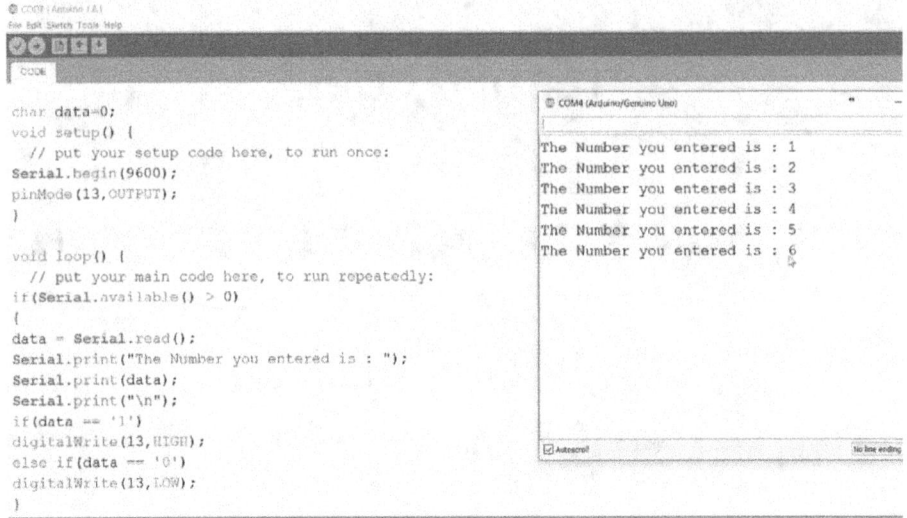

Now if we sent one as you can see or two or three or four. And the thing that we send here if we are using our let's say Android application and I think that we right here must come back to us. Now in order to see the changes here and when we enter but it must show us that number. You talked about applied to us and that one tool with that number that we liked. So if you wrote one it will start with one. Now to see the changes. Let's add a sentence before this one saying that I'm Bob your unthawed is OK. Now let's verify our code then upload it. Now if you wrote one word reply with a sentence the number you end up as one and the same for any other number. As you can see so it's I think this version of this sentence then it's sort of thinning the variable that it did that in this line. So it was built into it for us. We showed the T. As you can see this is basically how cynical medication work and this is test for our code.

INTRODUCTION TO SENSORS

We are talking about sensors Thighs is module six. Okay let's just take the content and the content we have is an introduction to sensor water sensor tibo sensor potentiometer PJ represent the temperature sensor and PIR sensor. We will be experimenting on the reason for on the last fall sensors and all other will be okay let's just come to the lead come to the first part our sensor can we sense a sensor can be said as anything that is taking as input from the environment and giving it output to the system which is connected to it which is used to change the change in the use is used to record the change. Then Vermont is cold the sensor will disorder technical limitation you can just simply a sensor is a machine that is connected to a system which is used to view any history changes in the environment to the machine it details it can be sent that event sensor is connected to the environment. If the temperature rises and it is a temperature sensor then the sensor will say that the temperature has been upward from is going up from 35 degrees Celsius to 45 degrees Celsius, it can be a humidity sensor which is saying that the humidity is increasing or decreasing with the time and but all these other sensors but the basic thing is that the sensors other devices that are taking the in the form data taking data in the form of inputs. Remember one thing that sensors are are only of two types. They are analogue sensors and digital sensors. In the real world or the physical world, all the signals are analogue but the computers have a built-in analogue to digital converters which convert the analogue signals into digital signals because computers cannot understand the analogue values. So, for the sensors, it is required that they can do only two values analogue values such as a multimeter which gives the value of a voltage such as 35 volts as 45 volts or 120 voltage. While a simple sensor such as a PIR sensor or a presence detection sensor can view only two values such as someone is in witness the value is high or one or someone is not present which gives the value low or you can say zero. Thighs were the basic thing or the sensor and there is a very high tech sensor. So, let us take about Wikipedia What does it say our sensor? Well, Wikipedia has a very technical definition of the sensor that it is a part of a subsystem that is used to process the data and yield the value of the environmental change to the machine. Well the sensors can be really simple such as a simple transistor for a temperature sensor or can be extremely Complicated such as M camera or a LIDAR sensor, okay thighs were the definition for the sensor alleges see, there are different type of sensor I have made a simple chart for your understanding. Here you can see there all the values from zero to 1023 410 words in encoders remember when things encoder can be very high they can be 32 bit 64 bit, but for our sake of convenience we are using 10 good encoders because it is present in Arduino, it can be even very high it can be to raise to 22 raised to 32-bit encoder for very fine and very precise knowledge for the sensor input and outputs okay. For the input values it can be 500 it can be 750 it can be zero it can be anything from zero to one zero to three so it gives a physical precise measured value. While for a sensor for a digital sensor. It can be only zero or it can be only a one there are no other possible instances. So since are of two types they are analogue sensors and digital sensors. For example, the camera that we use take those and take the values of the physical world in the form of digital signal incentives to the computer. The camera was on inter it connection or ITC protocol we will be discussing in the later chapter, we saw the functional of the camera it has a sensor it has the camera is itself called a sensor because it has a photo.on it and the photo is the light and gives a grid value in forming a giant matrix and the matrix is the values for the colour it can be RGB, red, green, blue and taking the value from it gives the image of the picture on the screen. Okay, about the sensors all sensors are input and there are some examples such as solar panels. You say that solar panel is a device that is used for generating electricity was the first and the most important use of solar panel is to detect light in heavy sensor where it requires the precise values they're using for the light values. Always Solar Wind allowed us because they give a very precise value and sensors are also have to type first sensors are of an active type and the second arrow passive type. In the active type, you can see that a sensor that does not require voltages or currents to run them is called an active sensor. So, the solar panel because it does not require any external power source to run it, while a passive sensor you can say if you can say a sensor is another example of that you can say temperature sensor transistor, such as lm 30. Well requires five-volt input for it's working. So, they are called passive sensors because they are taking power. So, active sensors are

Woodford most of the sensors that we will encounter in the physical world are always passive sensors because they have very precise value and they do not depend on the external conditions for their voltages so they give a very precise when Okay, about the leader. a leader is a type of sensor that is used to emit as light data is used to or well basically the function of the leader is to measure the distance. So, it has two parts, let me draw it for you. Here we are coming down just where it was again and you can see that here is an example of a leader. My drawing is not good sorry for that. a leader is like thighs rectangle box. It's not a triangle, I know it had two lights or you can say two devices, the first function is to emit light and that will collide to some object and then it will come back to another part. So that it can read that light is coming and taking a time interval in the processing unit. It is taking the pulse width modulation of the sensor there the pulse that has been sent as you know degree receives 30 degrees so there isn't small algorithm which says that for 30 degrees, the distances for Define meter and which by thighs method it gives a value to another microcontroller computer to which it is connected. Thighs were the theory. So, I have to tell you that all the sensors are input sensor but there is some type of sensor There are also such as output sensors such as leaders because these leaders give the password, first of all, they emit the right so, they give a value as output and when the light strikes to a solid surface, it comes back to the sensor. So, the sensor rated one again. So, thighs time it became an active sensor with a theory that it emits light such as an in the form of output and read the light in the form of input and that gives the value to the microcontroller process it with the help of an advanced algorithm and give the value of the distance. These are very high tech sensors because they are also able to calculate the speed of light. A simple example of a leader sensor is the ultrasonic sensor very cheap if you want to do extremely well. The sonic sensor is not a LIDAR sensor, but it works on a similar principle. Let me explain it to you. Thighs is an ultrasonic sensor you can see here it has the same functionality I saw earlier, but it uses ultrasonic waves, thighs point emits the ultrasonic waves and when it strikes to a substance it will be read with thighs module. And thighs both are doing there is a chip behind them that uses the algorithm to compare the time interval between them to find the distance because of thighs because it's very cheap because the speed of ultrasound is very long something about

360 to 400 meters per second. So, it is very easy for a simple microcontroller or simple processing chip to calculate the distance so ultrasonic sensors are very cheap, but there are very high-end ultrasonic sensors also available and we will be discussing them above In the future, thighs were the example of an ultrasonic sensor that is used to measure the distance we have later but we can use ultrasonic sensors for a small distance up to four meters. While LED is used to measure the distance from a 15 centimetre to about 18 meters. Well, they say it's 40 meters, but if you had them the sensor can give the value of fluidity meters also. And there is also some high tech leader sensor that can raise the value to 30 kilometres. They are used by most militaries for their operations to measure the distance between the mountains and other positions to give them the value for their post and other their other physical values. Okay, thighs were the theory of the sensor.

We are here at model six in Moodle six we are discussing water sensors and how sensors are useful in Arduino and other environments. Okay, first of all, let's get to the content is as follows an introduction to the sensor, what is the sensor then after we will be knowing about the type of the sensor that we use about the potentiometer will potentiometer is not a sensor but it is the most common thing to get to understand thighs or it is a water sensor. Then we will be talking about the electric sensor. Then we are talking about the temperature sensor and then we'll be talking about the P IR or passive infrared radar sensor fassi we will be talking about the passive infrared sensor. Okay, let's get started. Introduction Under the sensor, a sensor is a device whose function is to detect the changes in the environment and give a specific output to the user or the device. Well, you can say that if the temperature is 37 degree and a heat sensor is placed in the room and if the temperature changes to 37 to 39 degrees, it will show that there is a rising temperature in here that is about two degrees well it is easier to talk about how does the microcontroller or any other computer system get them all of them enriching? Well, thighs are the method of interfacing the sensor with the system by the help of programming and with the help of GPIO pins that are placed on the microcontroller boards. Okay, well since Okay, and sensors our the sensors are I will say the sensors are added sensors are used in various technologies such as in robotics as in automated cars. It can be used in industries well they are mostly used in industries. For example, the camera which we are using here is also a type of sensors because it is changing the world It is recording something that changes environment it is giving feedback, okay but there are different types of sensors we have we can talk about the physiological sensors or temperature sensor humidity sensor or we can say about the IR sensor. But, we are dealing with we are going to deal with the most basic to the most Ross a person-centred here these sensors can be generally classified only on our two tests, they are analogue sensors or digital sensor. Analogue sensors are the sensor which is similar to a potentiometer or you can say that they give a measured value while the data sense another sensor that can give only two types of value that is one or zero it can give us only the status high or status low while an analogue sensor can give us the value quite different such as higher, 35 336 or 590 etc. In thighs chapter, we will be going to India After we are going to learn that in Arduino there is only a possible analogue to digital converter that has a range of zero to two raised to 10 minus one or you can say that 02 to 1023. So, the range of the sensor is set like that as well so, that it can get read about our another type of the only another classification of the sensor is called as active classification or passive sensor can be an active sensor or a sensor can be a passive sensor, the active sensor is the sensors that do not require an external power supply to operate them. the example you can say a solar cell so listen can we use a very fine sensor for detecting the results of light in a room it is also the useful transmission of audio signals from one place to another I will be telling us about a position that uses sensors that need external power, such as a temperature sensor, a temperature sensor is a device would need power as an input. Well, here you can see that thighs are a temperature sensor there Reduce the value in the form of measurements you can measure the temperature outside it by using a library which is specially dedicated to the thighs temperature sensor and the library the main function of the library is to calculate all the environment variables and use it and suitable and take them the main function of the library to take the analogue value give it as input for the function multiply or do the mathematical calculation out of it and give the value output as the exact value for the temperature Awesome. Well, these are the analogue sensors while the data centre can be very simple, which are can be seen in the case as a p IR or passive infrared sensor. a passive infrared sensor is using for the right for reading the presence of a human in the range where they do not need x the value of the present such as they cannot be given they cannot give output as the measurement so they give the value in the form of ones or zeros or you can say that low signal if no human is present in signal high if the human is present. Okay well on Wikipedia there is a very beautiful definition of the sensor is given where you can read that a sensor is a diverse whose function is to detect the change in the environment and given us output to the microcontroller or the system in which it is connected.

It can give a value of zero to one or it can give analogue value. Well, the sensor of different type we are we know today here is some. So, the sensor data we know today is they detect light they can detect temperature they can detect humidity, they can even detect the pressure. Moreover, there are it is saying that there is always a sensor for everything it means that you can use a sensor or a group of a sensor to make some new type of sensor or new dyno device. Sensors have a basic application in many fields such as machinery, aeroplane, aerospace, car, medicine, robotics, and all other days to day lives.

Even our mobile phone contains two to three sensors such as the aeroscope accelerometer and there is a small sensor at the camera that detects the presence of human ear for that for giving a better performance of the car audio quality basically sensor can be anything in you can read here you can see here thighs is an infrared sensor when a light emitter object come in, thighs is an infrared sensor when an object comes in front of it, it gives a value high value in object there is no while with a while well, while will while when there is no object in front of it, he gives a value flow you can say that thighs are also sensor but thighs is an example of the digital sensor. It cannot do the distance of the measurement that has been taken here in the main sensor we are talking about we will be talking about both analogue and the digital sensor how to use them and the most important thing is there how to use them in logics how to get a specific type of logical requirement that we need here. How to get to perform a specific logical operation with the help of sensors such as running a motor brightening and led all thighs we're going to talk about them in further videos. Right now thighs were the introduction of the sensor. Thank you if you have any. Right now thighs were the introduction of the sensor. If you have any question please ask me or else Thanks for watching.

POTENTIOMETER

A potentiometer is a voltage dividing device but it can be also used as a sensor to measure the potential difference across the lines just talk about potentiometer. A potentiometer is a three-terminal device with a rotating the dial on it which works as a root which works as a potential divider. You can see that on the three legs the third leg is four-five worlds out input The middle leg is connected to the pin which is set as an analogue to set the voltage difference. Well, thighs are the basic pin diagram of the potentiometer it has an international symbol as thighs one in the Canadian and US symbol s as the exact line with an arrow lead just get a brief description about the sensors, Okay, on Wikipedia, there is a beautiful page which tells about the sensor it says that it is a three-terminal device with a rotating dial on it which can be used as a voltage divider as well as it can be used as a device to send the voltage difference across the terminals, okay. Here you can find different type of a potentiometer and thighs is the basic working of it. It is the fixed position it has two basic moving parts. First one is a fixed contact and the second one is a moving contact with a high resistive wire, high resistance wire placed on it when the dial has rotated the distance between the fixed terminal and the moving terminal increases and as the distance increases due to the slow the Potential potential and due to the ohms low the distance between the points and desert because of that, the potential difference between them decreases. Well, thighs were the basic working principle of the potentiometer and it condensed in nichrome wire for the resistive material while different types of sensors can be used here, which can use a fixed which uses graphite or other heavy materials for the resistivity element. Since Earth can be simply rotating there like thighs one you can see that here I am rotating the dial well but how to get it connected to the Arduino. I have shown you the previous video but I want to tell you here about the potential differences that have been created by the Arduino.

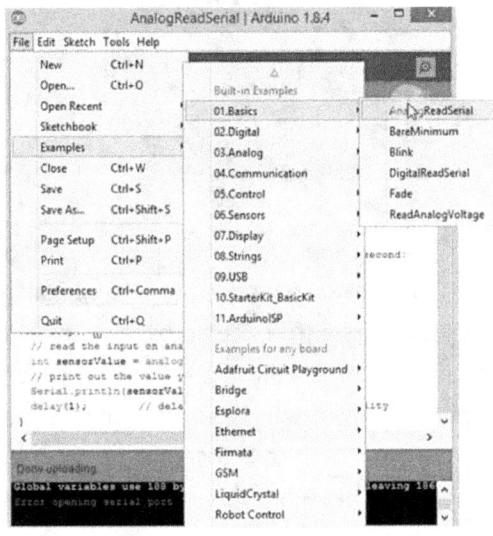

For that just go to our Arduino files, go to basics and select the first option that is analogue read serially in analogue see it's easier. You can see here in Word setup, it is calling up or it is calling Serial Monitor was that in the world room it is first calling the pin is zero.

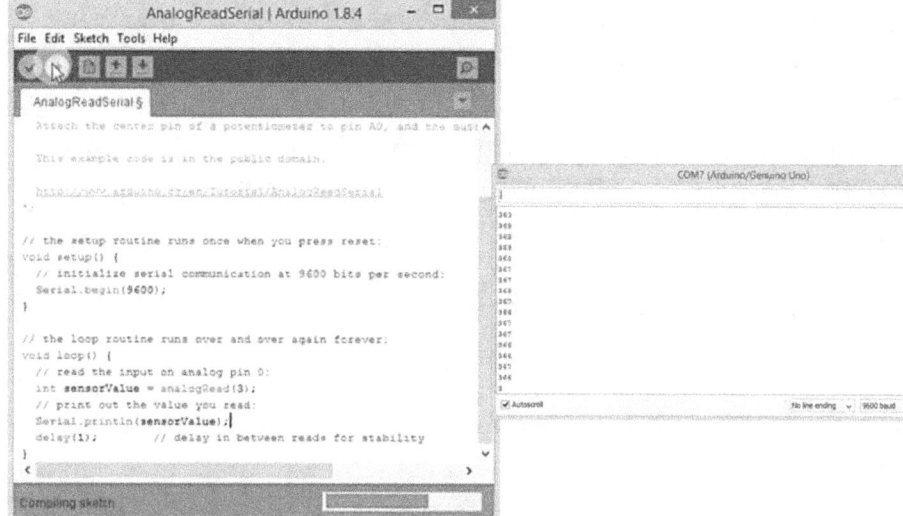

Well for my configuration I have connected it to the PIN three of the Arduino and in serial. In serial print, it is calling the sensor value or you can see that there what is the analogue to data's value that has been taken by the sensor ledgers uploaded to the world. It's taking some time and it's uploaded to the world. Now open the serial monitor. We opened it and here you can see that it's giving a value. Now just rotate the dial let me correct myself there is some wiring mistake. Okay, let's just get right. Okay, let's just experiment. First, open Arduino.

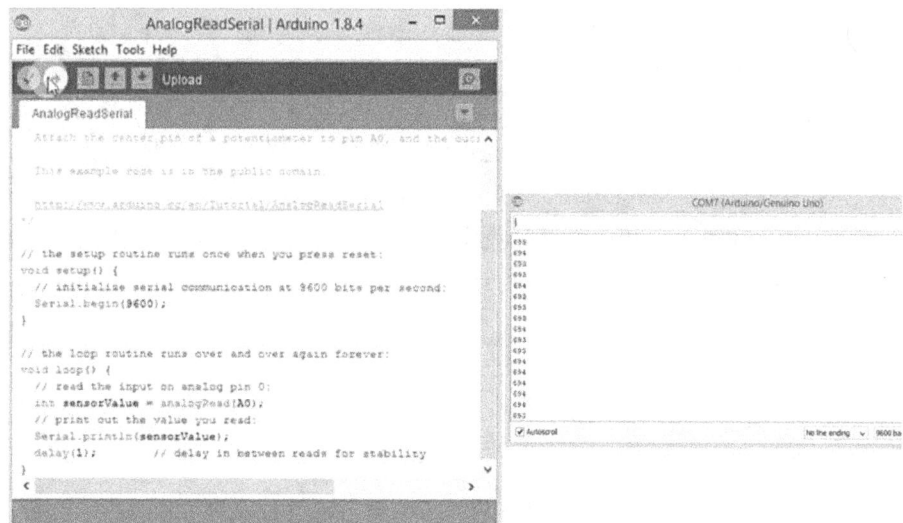

Go to files Let examples and go to basic select the first option that is data analogue that is analogue read serial, okay. In word setup, it has called the serial monitor with the word rate of 9600. And in worldview it is saying there the integer value, the teacher has given a sensor value variable visually storing the value of the integer and it is reading analogue rate is

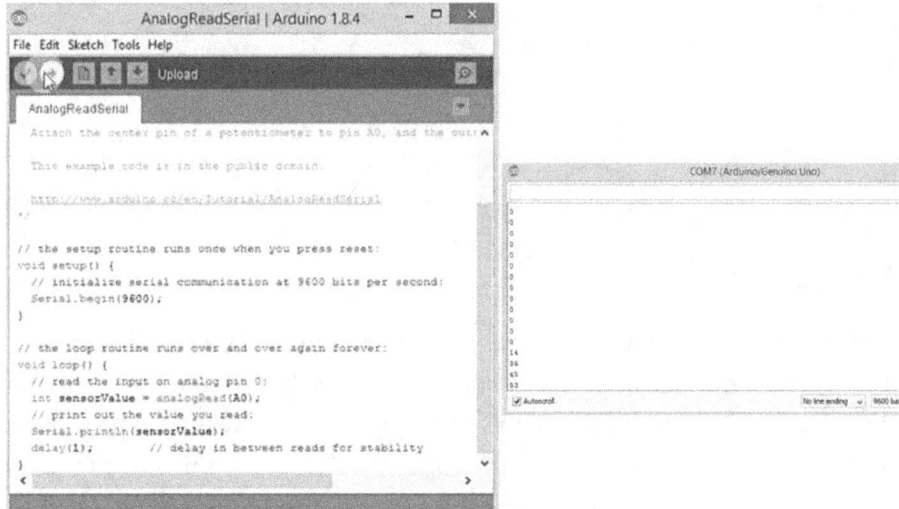

zero here it is reading the value and the potentiometer has been connected to one pin to ground one pin to positive or one to two analogue pins. After that, it will be printing the value as a variable sensor value will be showing the value of what the analogue grid has been done in with a delay of one millisecond. Okay, let's just upload it to the world. It's uploading taking some time and upload is done. Okay, let's just see the serial monitor and it's given a value of 693 and four allergists rotate the dial zero and it's going to 1023 Let me make a clear explanation of it. We are going to zero we are going to one zero to three let's just make it in between is given 671 Auto rotating a few times more, I'm getting 390 234 Okay.

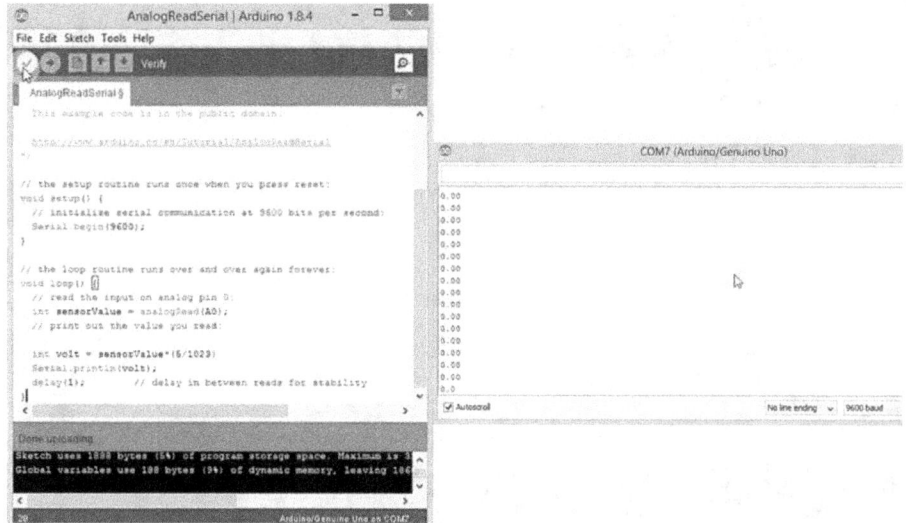

By thighs way, we can control the potentiometer and thighs is the method which is telling us what is the value of the analogue to digital converter. But if you want to get the potential out of it, there is a simple method just go here and you can say that to store a value in another go you're right an integer I nt we owe It It is our voter has been declaring and give voltage a calculation s e n s or VL essential value To clear the integer volt integrity value or sensor integer give it a multiply we are giving it multiply and then you do curly bracket write the formula five divided by 102301023 is the maximum value and five is the maximum voltage. so here the value of volts will be storing sensor multiply by multiply by in bracket five divided by 1023 okay right here voltages we hope O IT okay voltage is written here and it is taking the value within one millisecond. Okay, let's just compile it for checking. It's giving me an error. It is seeing me the voltage that is defined here. My mistake I am not terminated okay terminating it again giving it a compile mission compiling is done properly now uploaded to the world we have already successfully open the serial monitor and we can read the voltage that is zero turn the dial okay there is a mistake done by me that is the value of the voltage should not be an integer because it is taking my values that are out of decimal So, we should be giving it an FL o at float so that it can read the values in decimal okay uploaded to the world.

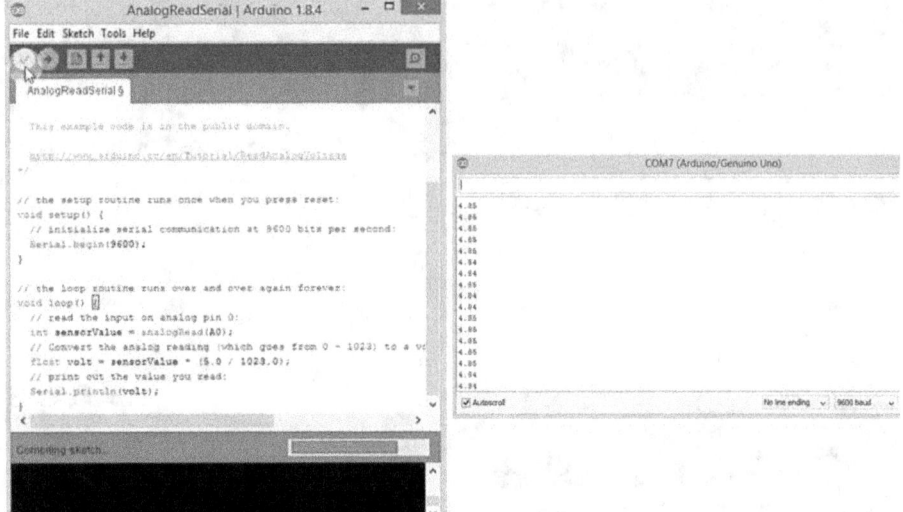

We are uploading it and now let us see the value. It's reading leading in double values. Okay, let's just come around there's some wiring mistake I'm just going to correct it within a second. Okay, let's just come here. First of all, let's see thighs index, first of all, we have called serial monitor invalid setup and invoice flow. We have taken pain to read the sensor value analogue value is zero. After that, we are given a formula that is for World sensor value that will be equal to five divided by one zero to three into a multiplication of sensor value, okay? It will, it will, after the calculation it will do the value for the voltage AC. Okay, let's try to compile it's clear no issue, you just upload it to the world. We have successfully uploaded it now just open the serial monitor and you can see value here. Now let's just change the value zero, it's giving me a voltage of 1143 2.45 or four, okay no issue giving a maximum value of five or 10 changes to something 316 system by thighs method a potentiometer can be used to read the voltage of cross the line well it is notice sensor but it is the best example to declare well it is the best example to explain about the analogue sensor values that have been GD that can that will be will that we will be reading in future, okay if you have any questions please ask me always thanks for watching

PIEZOELECTRIC SENSOR

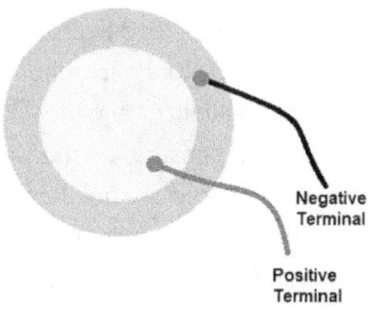

We are going to talk about PC electricity or PC elected sensor. Well PJ electric sensors are the special type of sensor when they are present they will generate a voltage Thighs is because of the electromechanical stress or strain that has been happening on the crystals. So it generates a voltage across the two plates that are connected to it. In here you can see that thighs are a PJ electric sensor and here is the first plate in between there is a material and outer plate is there which has been soldered by me for the future. Electricity is very important for defining for the one reason electricity is a very important indicator for using in sensors because it is used to generate pressure because it is used to measure or senses the pressure difference or you can say the rapid forces that are applied to anything, please electricity hasn't special effects. Visual electricity has been used in spatial sensors such as load sensor, because as the load is applied on the sensors because of the strain that because of the mechanical strain the voltage reading across the load changes, so, they are useful in many conditions. Here you can see that in the diagram when the military centre is stressed a voltage is generated. Let us see simple documentation on offer. Let us see single documentation. Let us see simple documentation of it at Wikipedia. I have found something it is saying that when the visual electric sensors are applied when the force is applied on the facility sensor Because of the crystal structure of the material that is being used inside it, it can be generally a ceramic or another crystal substance, when the forces applied on it, because of the mechanical stress the internal structure of the substance changes and thighs creates a potential difference across the surface across surfaces and media. And thighs generates a voltage difference between electricity city and visual electric sensors had been used in many applications such as products such as increased electric microphones, generation of electrical frequency, making microbalances and most important thing of visibility sensor is used in ultrasonic sensors. Let me show you. He had you see that these sensors they are all based on the electricity from they are all based on electricity you can see there the cylindrical object has inside of highs electrical component When electricity is applied or it gives the viewer superior electric effect to ask when the electricity supplied on the centre mix is generating vibrations and these vibrations are used to generate ultrasound waves as the same as thighs or the other side you can see there is also a cylinder when the frequency gets when the frequency starts on substance and reflects the centre due to the frequency generator it only generates voltage and by the by calculating the speed of the ultrasonic wave and the time difference between them It can detect the distances Thighs is a very important sensor here. Now, just wait a second. Now we are going to experiment on physiological sensors. Well, there are two methods to get the visuality sensor first is by reading a digital value but there is a problem because when I connected the visual interest sensor to a multimeter it is already generating some garbage values. And thighs can cause an error in the reading of electric sensor with the method of detail So, I have just used a method of analogue read because in analogue read you can get a clear value of measurement of what is going across it and you can get a value of voltage just go to a file. Select example, go to basics, select analogue read voltages, I have selected it Just a minute. Okay, we are connected the second we are connecting two pins first pin to the ground and another pin to analogue pin is zero you can connect anywhere to anywhere but let me correct it I am connected first in black one to ground and a red one to analogue zero. Wait a second. Well, I have connected it.

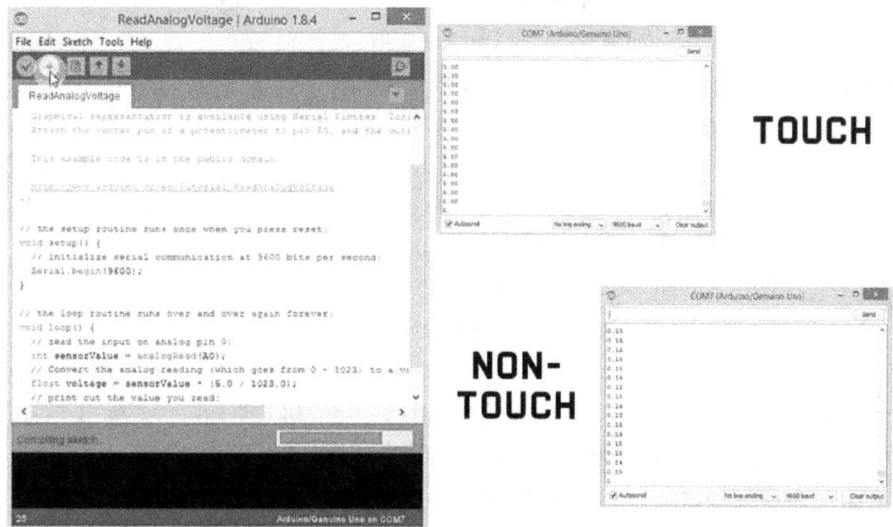

TOUCH

NON-TOUCH

Okay, I'm going to upload the code to it. And the code is uploading, there is a problem on the board. So I reconnected it. I am uploading the code. It's uploading and you can see that it has been uploaded, open the serial monitor. And you can see the garbage values that have been generated on the sensor. Okay, now go to the sensor and keep your finger on it and press it. You can see the voltage. When I press it, it generates a voltage out of five volts. Again, you can see it, it's given some garbled values it is getting down by a second. After that the mechanical force has been displaced completely within a few seconds, the value of the voltage goes back to zero. Again, I'm going to press it to show you whenever I press it, it generates a lot of stray voltage or you can see that it generates a potential difference across the splits which can be sensed by The Guardian. Okay, you can see here I am going again to present I'm again going to press it oh there you can see here whenever I press it goes to restrict fivefold okay by thighs method we can use the visual interest sensor in many devices and we can use it to perform another logical operation by the help of microcontroller we can define a logical program to read and we can get another function from it such as whenever someone plays some whenever someone applies the food on the sensor, you can connect it to a microcontroller and set is to make a buzzer. We know someone presses it will the microcontroller which will be the microcontroller will order the buzzer to sound so whenever is entered whenever someone is entering your room whenever he presses the sensor, the buzzer will read aloud song by thighs method by thighs musical operation which is written by a program that is written on Arduino. If you want to do it, I am going to check em jr function just wait a second. I'm connecting the led to the Arduino to make a small function of it. Here we go in we are going to perform a logical operation with the help of a preservative sensor as an input to the microcontroller will perform logic and a LED is connected to two to give it an output. Okay, we are setting says that whenever the voltage across the roof is a liquid sensors goals are high about three voltages the LED at the PINtwo will blink okay we are going to write the program. See the screen we are going to write it here. First of all, define the pin of the sensor slip in mode just to make connected p I n m or DPINM for the EP more see p number two as output Give yourself a number two output come down here give a simple if statement evil says voltage V o l TAG voltage greater than three give the statement d GITL w RI ta digital write PINtwo is will be high h l h you can see that, but it should be high for a few seconds So, that we can see it say delay, say 100 milliseconds it will be making the program much more clear. Okay, just upload it to the board. It's compiling and uploading.

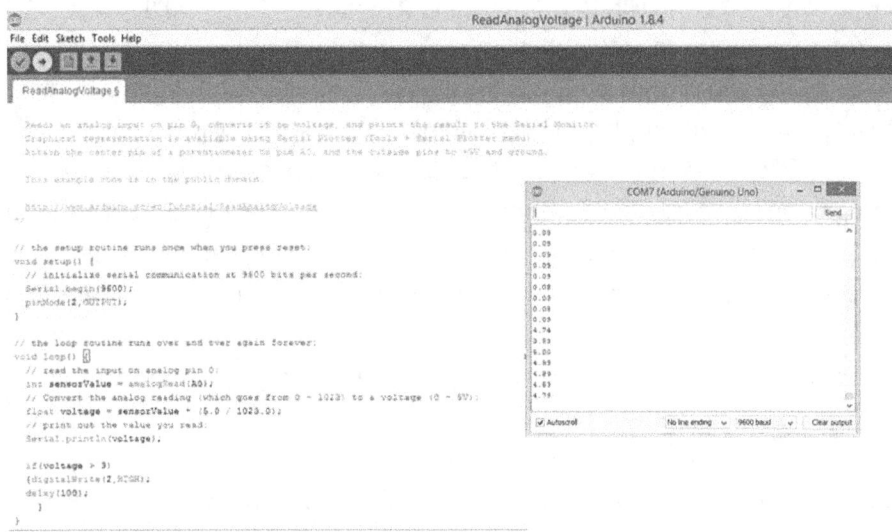

Well, the upload has been done successfully. Go to the serial monitor so that we can get a two-year option. Okay, you can see that it's about three words.

It's giving output has led High, again we are waiting for it to go down just wait for a second it's going down wait for a second we can see okay, let's just change the value from three to five it will be giving us a break better precision of the system for the system it is it will be giving us a better precision value. Okay, let's just see it whenever I press it saying above I will it is not possible, say about 4.5 it's just at thighs float so it can read 4.5 up, apply for the program on the board.

Okay, just open the serial monitor. And as you can see that whenever the five volts happens, it begins to flow. Just wait for it to go down. It will take hundred milliseconds Stop oh we have not put the statement okay go here, right e s e Elsa you two brackets right di gi t LWRIT because note writing the statement is becoming high for the permanent we have to make its laws. So we are going to make it law writers to Israel to the law in the statement uploaded to the world and now see it. Uploading program waits for a second okay go here in the serial monitor. Again you can see whenever it will go below 4.5 it will begin to stop it. The LED will stop. You can see it here once more. Let's just do it for a second is going down and you can see those systems By thighs way, you can use a physic electric sensor to perform an activity and write a simple program of it or to get an operation such as led high or you can also place a buzzer in place of it to giving you more of a visual song, a vision to give you a sound alert, okay, I'm going to save thighs code and you can find the code with the PDF of the document.

TEMPERATURE SENSOR

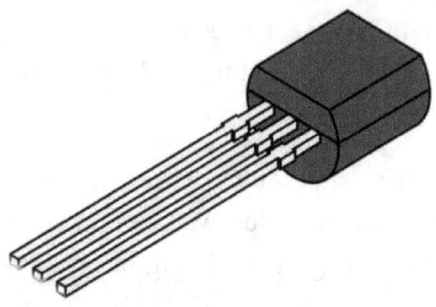

What is a temperature sensor It is a device that is used to measure the temperature across the environment or the object mutates in contact with it and give a voltage difference across the terminals of the Arduino? The voltage difference here is also given in the form of analogue signals because for the precise voltage it is not possible to take the signal in for more details. So, we are taking the sensor reading in the form of analogue signals Okay. Let's just talk about the temperature sensor here we have connected here the temperature sensor name as lm 35 is made by Texas Instruments and it is Earth transistor and which water linearity is means you can say that we know the voltage is applied across it will be swing low whenever a voltage is applied across it and whenever a temperature is increased across the frame of the sensor it will give a voltage difference occurrence across highs middle terminal. And we have connected the middle terminals to the analogue route of Arduino. So It will give a temperature difference and by having a small calculation you can get that there is a temperature difference written in the atmosphere or you can get the temperature of the environment.

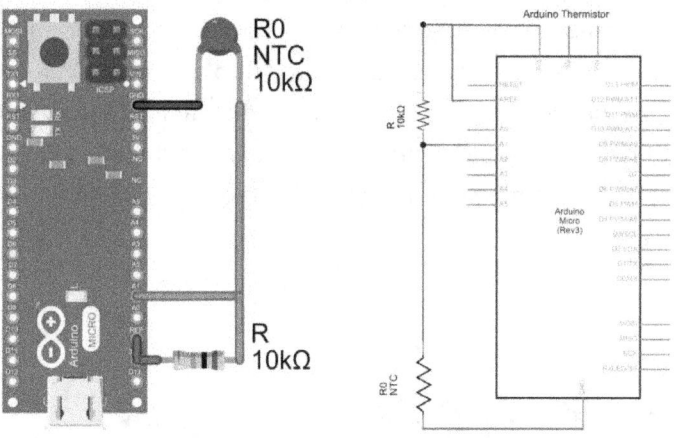

Okay we legislature water sensor, here is a data sheet I will be giving you the data sheet along with the PDF of the course. Okay, let's see, well, first of all about the datasheet is a special type of document that is prepared for the that is prepared with the datasheet a special type of document that gives the complete description of the electronic components about these physical dimensions is mechanical strength stress capabilities and the maximum voltage or current it can take. All thighs data are given in a specific document and that document is known as a data sheet. Here you can see that I have taken a datasheet from Texas Instruments about the LM 35 sensor, it is a linear voltage device and you can see that here are the physical damage dimensions And we are talking about an LP package three to 92 thighs is the centre that we are using here Okay, these are some mechanical stress extend capabilities its mechanical parameters as well as electrical parameters and working differences at other temperatures. You can see that here it has been given operations at the pin sir and how to connect it all the features are given in the datasheet has been being given by the company that had made the sensor they did gives the electronic engineer a complete knowledge about the sensor or the electronic component that he is using and how to use it or how to get the information out of it. So, a datasheet reading is very important for the engineers to get a complete brief knowledge of the sensor or the electronic device. Okay, here it's working temperature is minus 55 degrees Celsius to 150 degrees Celsius. Okay, let's just try to go To get a voltage reading, it's simply open the Arduino board go to Edit, go to examples and go to analogue read voltage for getting the voltage difference to you, we have connected it to analogue pin is zero and wiring is as per the PPT, you can see that is the first animal is connected to five words the last terminal is connected to ground in the middle terminal is connected to analogue pin zeros registered in the Arduino voltage sketch uploaded to the world.

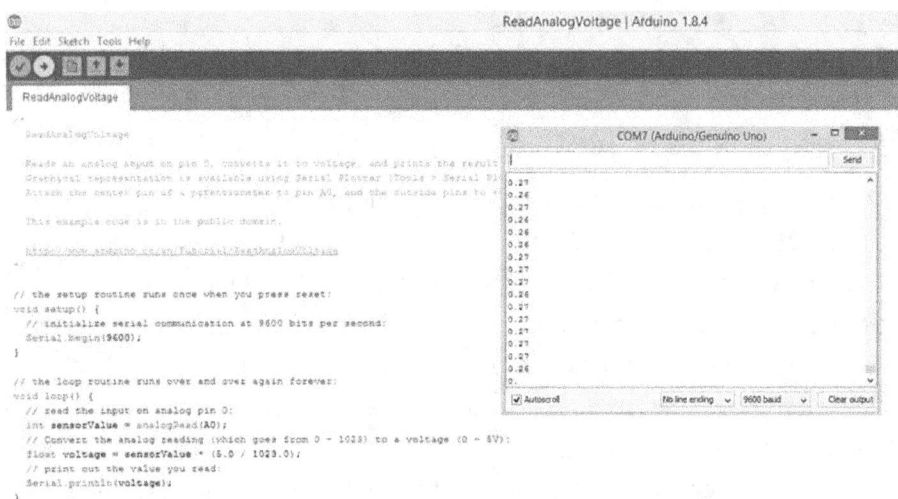

It's taking some time and it is taking some time it's uploaded on the board. Okay, let's see what is the voltage across it present it is 23 or 24. Okay when I test the sensor, the voltage changes continuously as I am touching because my hands are warm when I whenever I touch the centre is going to increase temperature. So, as for the temperature linear coefficient the value of the voltage across the terminal increases. So, the analogues read can and read high temperatures across okay that is for the reading of the voltage or how can we read the temperature I have found a beautiful quote here, here you can see that it is exactly similar to analogue read well we are going to do it in analogue read also multiply the value Okay. See here, thighs is the code that I have written first of all they are defining an integer. First of all, we are going to define an integer as v l for value in serial v 9600. For the operations, analogue well-read is zero because thighs are the pin we are connected for the sensor and it is the equation down here. For the value m v, we are going to well divided by one zero to four into 5004 c l we are going to divide MV by 10. Okay for Celsius and Fahrenheit You can multiply the CL into nine divided by five and four to do okay, just get the value for CL here. I'm going to upload it to the microcontroller and we are getting a difference of 100. For better understanding or quick operations, it's uploading it for a second. It's uploaded successfully in the world. Now, you can see the temperature and whenever I touch it, the temperature changes. I am going to do a practical operation. I'm going to put a mystic matchstick around it. Wait for one second.

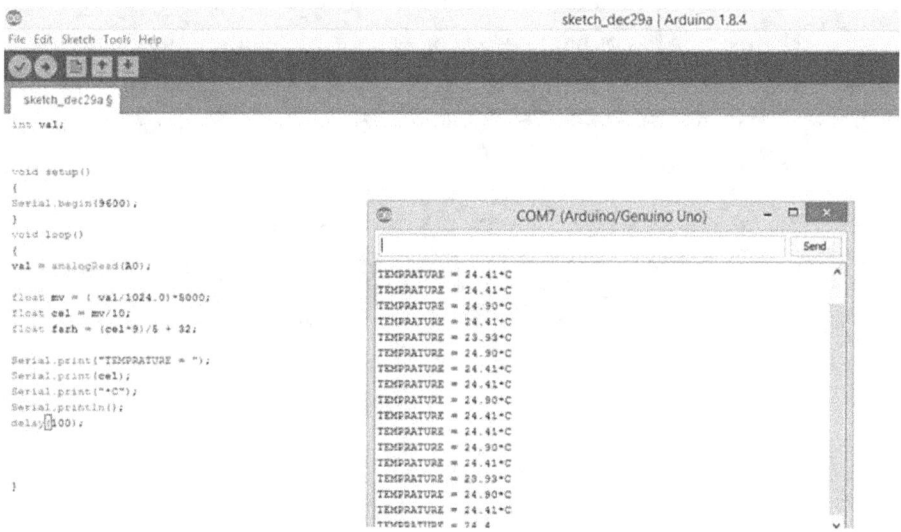

You can see that the temperature is rising continuously. Please be careful while doing the operation because it can fry the sensor if you have taken the message. Longer times. And for reading thighs Fahrenheit, let's just do a small change in the code where the cel is written go there and write f A r h for Fahrenheit because a Fahrenheit here is the variable that's true the value for Fahrenheit as for the calculation of CES multiply by nine divided by five cluster tables, okay, let's just upload it to the world. And okay, let's see. Yeah, you can see that the temperature is okay. Just a mistake. It's not I will be doing F for Fahrenheit, uploading it to the world. Okay, you can see that there is 8990 or 80 degrees Fahrenheit and whenever I touch it, the phonetic values changes. Okay, but what do we have to do a logical operation. I will be including logical operations from here for getting grip information or brief technical knowledge of the program and how to how does the program work? Okay go here in Word setup right p in just a minute right p l n import the EP mark, we are going to collect an LED at p number two and set it as output your status at output terminals the line come here and I am going to select the value c l for my understanding that is good. Go down write the if statement evil the Sl p l great greater than, say 32. The digital right LED will be high D, just a second, the ID it l w l t w capital as for the syntax to and it's important to keep it Also SNS statement because without s you cannot turn out you cannot turn off the LED use statement right delete all right, two comma L o w okay terminate the line and just upload it to the word wait for a second I am going to connect an LED okay here you can see that I have connected an LED to led pin five just see here Okay, I have connected an LED to led pin to okay we are going to upload the code and give it an MST MP at 80 you are etc and so on.

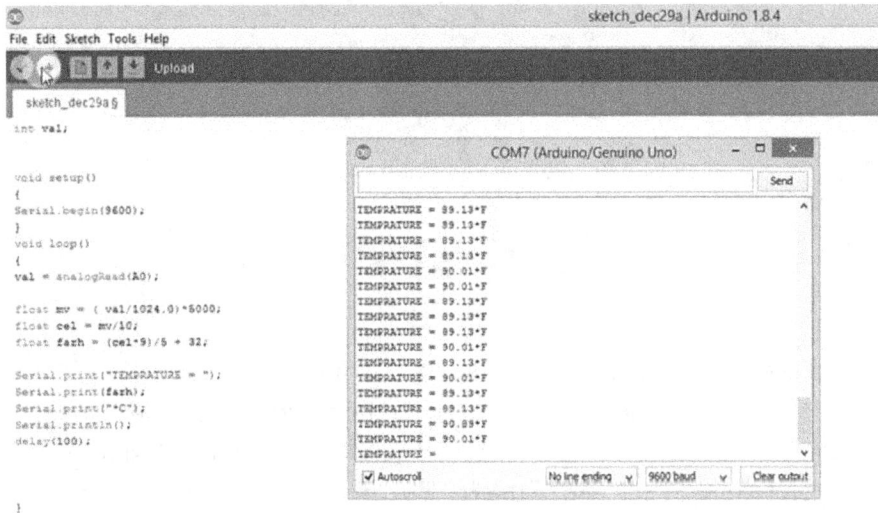

You can find the code at the beginning of the module for your practical purposes, or if you want to experiment, I have made a small area the peace capital and we want the fee to be small. Okay. Now, product two awards it's uploading just for a second. Okay, the upload has been successful. Okay go here and let's see the temperature is 23 degrees Celsius. Whenever I am putting my hand on the sensor the temperature keeps rising and it whenever it rises above 32 and goombah changes from 32 to 34 quick operations. Okay, we are set to 230. Again we are going to upload it and the state is being uploaded. Now, open the serial monitor once again and I'm putting my hand on the sensor so the temperature becomes high and let's just make it about 29 Now let's see well you can it is a better practice to set the values as per your precision because it will help you in many sensor calibrations I have calibrated it to 29 degrees Celsius. The word Fs return here was in text mistakes, I will be correcting it. There are two mistakes here. First of all, I am going to write C for thighs and the light will be going for the next hundred millisecond to keep the program stable and give us a reasonable operation. I am uploading it you can see here it's uploaded. Wait for a second. Open the monitor. Now. Correct your area just a minute there are some losers I have corrected it now whenever I'm going to put my fingers across the temperature sensors, the value of the temperature increases as follows. And whenever it goes about 29 degrees, the light will glow in when I put my hand back, the temperature will soon decrease and the light will turn off just wait for a second. It's going to 29 and it's flickering because it's oscillating between 28 and 29. Now you can see that the light has been turned off again. The light has been turned on for a few seconds after getting below 29 it will turn off okay, you can see, here again, thighs was the LM 35, Texas Instruments answer.

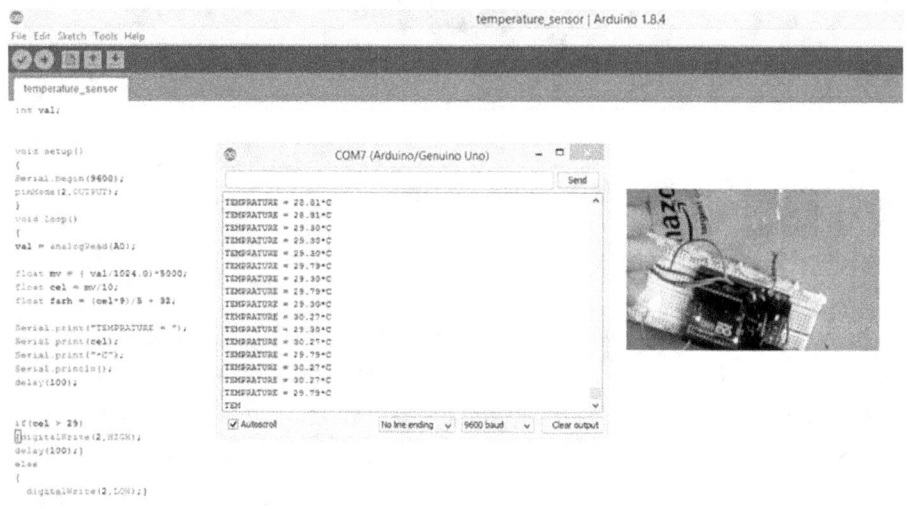

Thighs were well thighs was lm 35 temperature sensor made by Texas Instruments and in thighs program in thighs in thighs module we have talked about the temperature sensor and what is the datasheet and how to read the datasheet for the temperature sensor we have a small program for reading the voltage across the Tim SR sensor and as well as how to read the temperature across a temperature sensor it had formed a logical operation. See once again it's just going to blink and whenever the temperature goes below 29 degrees it will turn out well thighs was the operation performed by to give to get it to give you a simple logical understanding of the program.

PASSIVE INFRARED SENSOR

We are going to talk about the passive infrared sensor or you can say the IR sensor PS I the P IR sensor is a special type of sensor that is used To detect the presence of human within a range of 10 meters, it sensors work on the basic principle of detecting passive infrared waves that are coming outside the human body. Well, you have seen that the human voting Have you have seen that the human body emits some radiation that is in form of infrared and that is invisible to us, but when the sensor here you can see there when the sensor here takes the infrared radiation on it, it gives a value high to the microcontroller. Well thighs are the sensor you can see here and thighs is a Fresnel by and thighs is a Fresnel prism, let me give a current vision of you. And thigh is a professional prism that is connected on the sensor so that it can gather infrared radiation from all the 360-degree directions. Okay. Yeah, the thigh is a complete model of the IR sensors. Okay, let me just tell you about yours inside there are some sensors that are available that can detect A 200 meter of Denver that are not in our course. So, we are talking out of yours and that is very easily available and it can detect the human problem within a 1010 meter range. Its sensor is used to perform various activities in view and various give and give various logical operations as an opening of gates and even did it in the presence of human to give an alarm to the person who is watching it. Okay, let's just go to Arduino and write a program for it. First of all, in Arduino, you have to say pin mode to command output. corrected PDU output again to is printed output and we want to view in a serial monitor so serial dot veg I m serial dot begin is there in 9600. Okay, no problem. Now in here, we need an integer, two different Okay we will be defining it as a global I nt D for detecting equally to zero initially Okay, come here say D is equal to d IGITL r e a d d it'll read to okay we are reading data into and storing its value in interior D, okay? Nows e r I l.pr INTLNPL en la, we are reading the tensor two in the variable a loose code in variable D so let's just a kid was given a dealer for better operation of the system say 10 milliseconds Okay, now just connect the Arduino board connected you can see the sensor here sensor is there, uploaded right CPS in Sussex It's uploading it's just taking some time it was just uploaded and it's uploaded. First of all, I'm taking my hand in front of the sensor.

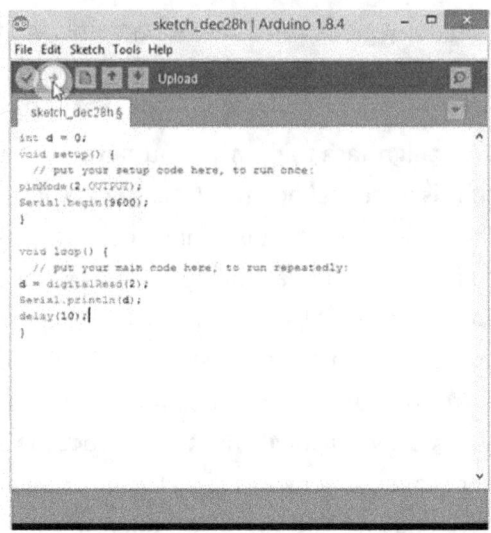

Okay, just a minute I lost Arduino, go to Arduino open CL monitor and you can see the value is low. Okay the sensor is not settled here are two variable registers or you can say potentiometers for correcting the sensor, just dial them to high sensitivity and another respond Delia give them high value. Now you can see there the way you can see here the video sensor is going high because my hand is placed over well I have said thighs density of the sensor high so that it is reading ping me I am sitting here on the table and it is detecting if we set thighs value of the sensor too low, it will be very difficult for it to detect me and I have at present status its value to 10 meters and the value set high for the dealer so that you can see that a sensor is giving me a high value it means that the sensor is saying that someone is the age present someone is present in the room. So by thighs method, we can use it to detect variable very. by thighs method, we can use it to detect humans and another type of life forms such as kettles, dogs, etc. Thighs are a very useful sensor for the opening of gates in the shopping mall universe in their automatic gates. They are working on the principle of Pac.

SERIAL COMMUNICATION INTRODUCTION

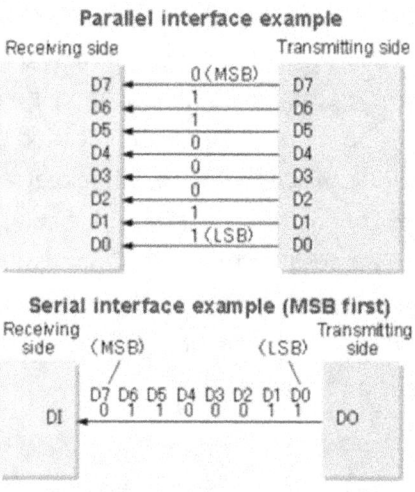

Serial communication basics and what are the contents here you can see that the contents are as follows an introduction to serial communication, serial begin details serial wire writer and serial reader and we will be talking about all these topics in the later videos. First of all, we are going to know about what is serial communication so let's get to its introduction about serial communication is a method of communication between two computers. It was the first design with the help of Rs 232 cable there is a port in your computer there. It has a cable which is in blue and has multiples holes in it. Thighs cable can is used to connect the monitor to use the CPU and thighs cable and also connect a laptop to another monitor. Thighs cable is known as rs 232. Thighs cable runs on the protocol name as the serial communication okay our serial communication and there is also a similar way which is known as parallel communication. Here you can see that about in parallel communication there are multiple wires connected to the between the two chips together to process the data in a very high rate high-speed rate while in a serial communication there is only one wire and physically there is two-wire to communicate between them. First one is the clock and the second is the data line. You can see that first data D zero enters another CPU then Done D two dental three and the series going on Buddhism. So, it is called serial communication because each breed is entered in another computer from the host computer in a continuous series. Okay. In Arduino, thighs are the most common method for connecting Arduino to the computer. You can see that here thighs is a cable in which I have connected it via USB to the Arduino board. Thighs Thighrd is communication. Thighs method of communication is also serial communication. Well about the serial communication Let me open the Arduino in Arduino. You can see that here, whenever we want to let an open whenever we want to connect the Arduino to the computer we have to write str I dot v eg in, which means that the serial Commission's comm does it Thighs means that the sale commissions will be started okay due to curly braces and here we write aboard the board rate is the rate at which the data between the two machines is transferred.

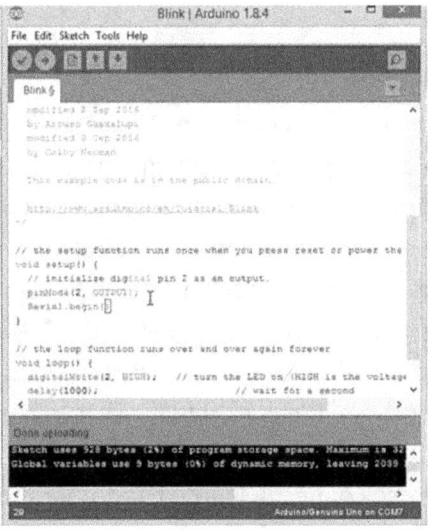

It is said here are standard values for the border. Let's see here 9600 192 double 038400115200 and so will the baud rate for both the computer need to be same here you can see there are no borders between them is 9600.

So you can get all the values and you can get the data in the room. They will form but if the board has said the value of 9600 in the Moodle year opening in 9219 200, it will give a false value it means that the system cannot read the data. So, for communication, it is necessary that both the computer asset to have a common board and there are other options such as newline and all thighs for the communication you can see here and let's just get rid to 9600 and you can see that the communication is done properly. To begin serial communication just write serial dot begin and give it a vote rate of 9600. I am going to do an experiment I am giving 115200 baud rate.

Okay, here I'm going to delete the program and writer capital S e r I I s e r l dot p RI NT Thighs means that the serial world has been printing on the computer and giving it a say hello h e l l o terminate the line and so on. Okay now uploaded to the world, first of all, I need to set a border delay function so that we can see the data end Why is the delay over 500 milliseconds that will be enough for us say 500 milliseconds,

terminated the line and again upload it to the board. Now open the serial monitor. What can you see you cannot see anything because I have said the voter ID and 9200 while in the system I have divided 1520 change the voter to 1500 115200. Now you can see that all the lines you can get the data in a readable format but you can see that it is not terminating continuously. It is all going in one line. So, for that to get the data every time a new line write the function serial dot print l n ln will introduce every time a new line of every bit of data has been having. Okay See here, you can see that the data is coming in every time in a new line. Thighs are the method of serial communication between the systems. You can do a lot of functions with the help of serial communication you can read the data you can write it with the help of a monitor here you can see that there is a space in here you can write the commands inside the Arduino board here we have not programmed anything in the Arduino to read the command so nothing will happen in the later videos we are going to write the functions in the study about how to read the command from the serial monitor when it is given from the computer serial monitor. Okay, now you can clear the output here. And let me show you something I have found a very beautiful document or a video on what is serial communication. Here the serial communication is the method of transmission of data Between the two computers or you can say to electronic devices as of now, it is a very simple method and it is more reliable in today's world than thighs parallel communication because today's world the speed of data sending is very high. So, it can match up to the older parallel communication methods must remember one thing more, but remember one thing more clearly in when we are talking about ultra high speed is there then you can say the CPU talking to a RAM, it is always parallel communication because there is a lot of data that should be gone between there should be a send and receive between the two systems for high-speed communication. So, the relationship between you can say, a CPU and RAM is a little communication via communication between us CPU and monitor is serial communication or you can say an Arduino and a computer are also doing the communication in the form of cereal and there is also pieces Ci and PCI Express cards that are using communication are an example of serial communication. Okay you can see here are the cables first of all serial communication was designed only to communicate the computers that are very far away you can say the distance between two states or two cities that was devised to communicate between them and for high distances the using of parallel communication become ultra-expensive. So, there are different methods of communication universal communication, you can see here, the first line is for data and there is also a second line that is for the timing clock between them to get their intervals done properly and gather data and same voter ID in both PCs. And there is also a word of the CL vs and that type of users. Serial versus parallel is a hot topic of discussion but remember that the ultra-high components when requires the very high speed of communication by low communication is preferred because it's simpler, simpler and head a lot of wire for high-speed communication between them while for long distances and Most common use or method of communication between the two electronic machines is to reseal communication. Now, remember one thing that in Arduino Uno there is only one port for serial communication as PIN zero and PIN one, while in another word such as Arduino Mega in Arduino, view there are three channels for serial communication such as communication one, serial communication two and serial communication. You can see here serial communication is the application and here you can see their serial communication can also be performed on other devices and Arduino such as insert name MTU 6050 and Arduino board communicate will work with each other with the help of serial communication and ITC protocol Also, you can see that in Arduino, mega and do they have three communication channels as RX TX one for serial communication one up to RX tx 347 communication Okay, let's just recall everything we are talking about serial communication Just a second we are talking about serial communication the difference between serial and parallel communication, what is the baud rate and how the serial communication between two happens it consists of two wires first for data and second for timing logging. Okay, we have made a small program here, I will begin showing you the program in the first section of the module you can get the program from there and perform your experiments say OK, you can perform your experiments there.

WHAT IS ULTRASONIC SENSOR

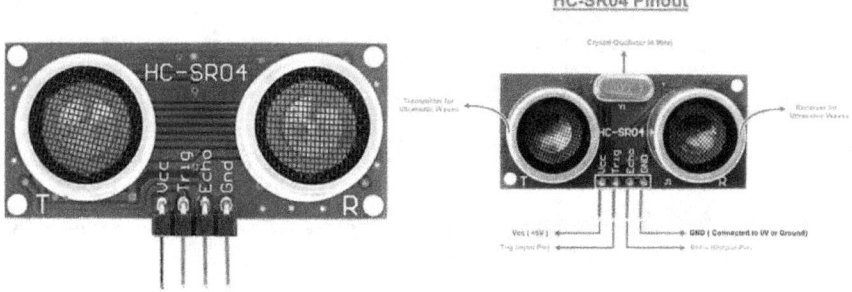

It measures this by sending out a sound wave at a specific frequency and listening for that sound wave to bounce back. By recording the elapsed time between the sound wave being generated and the sound wave bouncing back it is possible to calculate the distance between the sonar sensor and the object and instruction that's presented not short easy to remember for. As you can see this is all this is the ultrasonics And so this is the send-up. This is the receiver this Andersens ultrasonic wave. It hits the wall and returns to the sea of us so we can simply measure out of the distance between the sensor on the wall by using this simple equation. This stance would equal speed of sound multiplied by the time taken by divided by two. Since it is known that sound travels through the air at about three hundred forty-four The second you can't take the time for the sound wave to return and multiply that by three hundred forty-five. Forty-four meter to find the total round trip distance of the Soundwave that was generated by the ultrasonic since all round trip means that the sound wave travels two times that distance to the object before it was detected by the sensor. It includes that Treb from the sensor or from the sonar sensor to the object and the thread from the object to the ultrasonic sensor. After all the sound wave bounced off the object to find the distance to the object simply divide the round trip distance Imhoff which is clear in this simple equation. Distance would equal speed of sound which is 344 multiplied by the time taken. That we will measure divided by two because this is the first distance. This is the second distance. This is the reflected wave. This is the transmitted wave. So the distance from here to here must be the total distance divided by two. As you can see this is the uncle so and so on. This is the object now. This is everything about the ultrasonic sensor. Ultrasonic means that it sends an ultrasound wave. You can't hear the sound wave cause it's in the range above what human can hear. But they did they don't exist.

SOFTWARE REQUIREMENTS SCHEMATIC

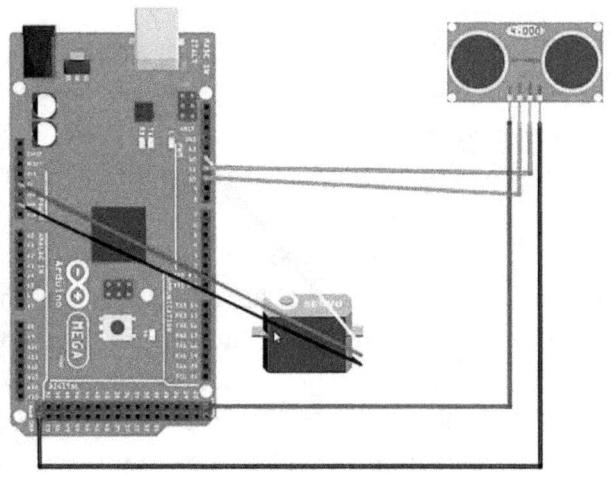

We will talk about software requirements Arduino code which will be installed on Arduino memory and will take control of our ultrasonics and so on several matters. The ultrasonics and so-called will measure the time that the wave takes to travel from the sender to the receiver. Divide that by two. We have that interface code for your computer that will throw a radar screen on you or the screen or your computer screen. And we have the schematic. Now let's look at the schematic. As you can see you can use Arduino on or maybe a gun depending on what's available for you. Here is the ultrasonics and sort of that we will use. And here is the set of all that we will use as you can see here. We connected two terminals with the settlement or two five vaults underground why the other terminal is connected to number 12 on the board-wide. That's also since it has two board terminals 1 4 5 volts and 1 for ground. Now there is another two terminals and ultrasonic sensors. This one is for. And for eco. They are connected to bin's number 10 and 11. Now I will teach you how to create a schematic for this project. You can do it using brought us. We did provide a course for simulating Arduino boards using brought us simulation without buying an audio board. You can check our profile and send us a message to send you a discount coupon to drawing that cost. But for this course, I will teach you a new software called frit Freethought Xining or fly zing. You can name it whatever you want. Just call me that name. Flooding. Then go to Google right. The name of that software as you can see it's an open-source hardware and intuitive that make electronically accessible as a creative material for everyone. A few collective will be directed to the company website as you can see. This is all on board. This is a big board and you can simply create it. Go to the download section and simply start downloading. Choose your operating system Windows. Click save and you will download will start. Now what. Once the donor is done you can simply get the zip file extract it and you will get this folder. Double click on that eaks E5 and the software will open up for you. Now just click the file and you as you can see this is a big bold you can move any item. Now let's start by then. I was bored. Go to my controllers Arduino then go on from our box of the set of gay let's call the seven months or two that I lean toward. Let's zoom in hands. The block wealth must be connected to ground that adware must be connected to a sort of Oddworld must be connected to 5 volts. Let's zoom out a little bit. Most connected he is to the 5:01 the first wild which is the feedback one when connected to being onboard let's say 10 or three. Now, this is all in the top 10. So we contacted the set or now we can connect the Aigues plus and extensor OK since ultrasonics and so doesn't have a library here. We will simply connect it just like in here. The board two wires 1 4 5 volts on fall down. Let's take off from five phones on connectivity. Let's take another while from the ground terminal connectivity is OK. Now and here we will take one line out from five Firefox and another well from the ground. Now we have two wells. Must connect them to bend number four. OK 10 11 12. As for the feedback from the 7.00 OK. Now let's add these two files. The first one is with number 10. This one is overthwart here. The second way of as with number 11 OK. Now we can simply connect ultrasonics since or here or another tear the breadboard and our ball will be good to go. This is only a demonstration of the schematic to show you where the fingers connected. You need to know that you can hook up a DC adapter here 5 volts or seven volts all you can simply Blug 5 volts on the ground here from our supply that's it for this lesson. If you have anything that is not clear and the schematic for the Arduino server.

TEMPERATURE AND HUMIDITY SENSOR

It also works with the H.T. 21 and DST 22. If you have these two as you can see this is how the sense of looks it has a plastic box a blue plastic box and three bends. Now before we store this image this ablaze the sensor values to the serial monitor as you can see it displays the humidity and temperature and the whole time. These are the zero communication window from Arduino id Software. They're guarding the schematic and how to connect this sensor to Arduino owner. As you can see Ben number one must be connected to 5 volts. As you can see this is the first spin. It's connected to 5 volts from the Arduino board. Bill Number two must be connected to number two. But number two is signal, Ben. So it must be connected to that Ben. Number two an Arduino board and it must be connected with 5 volts using a resistor. We choose number two because we are not a code that uses the number two. And the Arduino library board you can choose any other Ben if you want. The last pin is connected to that ground. Now before going any further depending on the sense that you will have you need to search for the schematic and how to connect it. We have the arch d 11 so we have to say as for the t 11 humidity and temperature sensor as you can see this is the schematic OK. We are waiting for the schematic to load and let's go back and see the schematic of the sensor as you can see it has four bends. The basic module will cost around $5. If you bought it alone it has a very small size it needs from three to five volt to about need. It will consume 2.5 million they're using from the water source during conversion. It has four bends. Now we can download its datasheet but this is the Chinese version so it won't tell us. But as you can see this is the connections first. It's connected to the 5 volts as we mentioned. Every other second bin is connected to the microcontroller which is, in this case, the board and it's connected to other more than 5 volts. The fourth bin is connected to the ground exactly as we mentioned earlier. And this schematic as you can see here. So that's it for the connection that the HD 11.

THE END

www.ingramcontent.com/pod-product-compliance
Lightning Source LLC
Chambersburg PA
CBHW052347220526
45465CB00003BA/994